SECOND EDITION

ONE WEEK LOAN

SECOND EDITION

INTERVIEWS

Learning the Craft of Qualitative Research Interviewing

Steinar Kvale

University of Aarhus

Svend Brinkmann

University of Aarhus

Los Angeles • London • New Delhi • Singapore

For information:

SAGE Publications, Inc.
2455 Teller Road
Thousand Oaks, California 91320
E-mail: order@sagepub.com

SAGE Publications India Pvt. Ltd.
B 1/I 1 Mohan Cooperative
  Industrial Area
Mathura Road, New Delhi 110 044
India

SAGE Publications Ltd.
1 Oliver's Yard
55 City Road
London EC1Y 1SP
United Kingdom

SAGE Publications Asia-Pacific
  Pte. Ltd.
33 Pekin Street #02-01
Far East Square
Singapore 048763

Printed in the United States of America

Library of Congress Cataloging-in-Publication Data

Kvale, Steinar, 1938–2008
InterViews: learning the craft of qualitative research interviewing / Steinar Kvale, Svend Brinkmann.—2nd ed.
    p. cm.
Includes bibliographical references and index.
ISBN 978-0-7619-2541-5 (cloth)
ISBN 978-0-7619-2542-2 (pbk.)
    1. Interviewing in sociology. 2. Interviewing. 3. Sociology—Research—Methodology. 4. Qualitative research. I. Brinkmann, Svend. II. Title.

HM48.K9 2009
301.072'3—dc22   2008003528

This book is printed on acid-free paper.

08   09   10   11   12   10   9   8   7   6   5   4   3   2   1

Acquisitions Editor:	Vicki Knight
Associate Editor:	Sean Connelly
Editorial Assistant:	Lauren Habib
Production Editor:	Astrid Virding
Copy Editor:	Jamie Robinson
Typesetter:	C&M Digitals (P) Ltd.
Proofreader:	Ellen Brink
Indexer:	Jeanne R. Busemeyer
Cover Designer:	Gail Buschman
Marketing Manager:	Stephanie Adams

CONTENTS

LIST OF BOXES

LIST OF TABLES

LIST OF FIGURES

———•·•◆•·•———

PREFACE

———•◆•———

This second edition of *InterViews* has a second author, Svend Brinkmann.

In this new edition, we have retained the basic contents and structure of the first edition (published in 1996) and also attempted to provide practical guidelines for novice interviewers on "how to do" interviewing research and to suggest—for novice and experienced interview researchers alike—conceptual frames of reference for "how to think about" interview research.

We have drawn in some of the multitude of literature on interviewing and qualitative research from the last decade. A variety of interview forms are addressed in a new chapter on interview forms, and, in contrast to the earlier emphasis on harmonious and empathetic interviews, we put more weight on active confrontational interviews. We have also given more attention to the linguistic aspects of interview knowledge; there is thus a new chapter on linguistic modes of interview analysis.

In the first edition, a tension of interviewing as either a method or a craft was pointed out, and in this second edition there is a stronger weight on learning interviewing as a craft. This involves an emphasis on learning from "best practice," and throughout the book we discuss examples from historical interview studies in the social sciences. We also suggest practical and conceptual learning tasks for the readers.

We have put more emphasis on how epistemological conceptions of the production of knowledge through interviews affect the many practical choices to be made throughout an interview investigation. Furthermore, we more comprehensively address interviewing as a social practice, embedded in an interview society. Thus the ethics of interviewing becomes a pervasive aspect of interviews. The ethical dimension of interview research is extended from ethical rules to encompass the broader fields of ethical and sociopolitical uncertainties in social research.

ACKNOWLEDGMENTS

Many colleagues and students have inspired us in the making of this second edition of InterViews. We would like to mention in particular the work of those at the Center of Qualitative Research in Aarhus—Claus Elmholdt, Lene Tanggaard, and our secretary, Lone Hansen. Lisbeth Grønborg, Merete Poulsen, Maria Virhøj Madsen, and Sidsel Carré have provided valuable assistance. Carsten Østerlund and his group of doctoral students also gave us important feedback on the manuscript. We would like to thank the following reviewers for many helpful comments:

Lisa Diamond, University of Utah

Kimberly Gregson, Ithaca College

Wilson R. Palacios, University of South Florida

Allison Tom, University of British Columbia

Marcus B. Weaver-Hightower, University of North Dakota

Artemus Ward, Northern Illinois University

We managed to finish the manuscript for this second edition of InterViews shortly before Steinar Kvale passed away in March 2008. Steinar was very pleased with the way that all the people at Sage Publications helped in finalizing the manuscript, and we wish to extend our thanks to all our editors.

—*Steinar Kvale*
Svend Brinkmann

INTRODUCTION

⬦

I f you want to know how people understand their world and their lives, why not talk with them? Conversation is a basic mode of human interaction. Human beings talk with each other; they interact, pose questions, and answer questions. Through conversations we get to know other people, learn about their experiences, feelings, attitudes, and the world they live in. In an interview conversation, the researcher asks about, and listens to, what people themselves tell about their lived world. The interviewer listens to their dreams, fears, and hopes; hears their views and opinions in their own words; and learns about their school and work situation, their family and social life.

This book invites the reader on a journey through the landscape of inter-view research. We shall outline paths that learners may follow on the way to their research goals, and we shall provide conceptual aids and toolboxes that may facilitate learning the craft of interviewing. In the conceptual chapters of the first part of the book, we put forth conceptual frames of reference for reflecting upon epistemological and ethical issues along the way. In the prac-tical chapters of the second part, we outline many of the specific tasks of interviewing, and also suggest some exercises for acquiring interview skills. We start with conceptualizing and designing an interview inquiry, then treat the interview situation and present different forms of interviewing, address the transcription from an oral interview interaction to written texts, and out-line a rich variety of analytic tools for the analysis of the interview texts, before concluding with a discussion of validating and reporting interview studies.

Three main perspectives on interviewing as research go through this book. We treat interviewing as a craft and emphasize the craftsmanship of the interview researcher. We understand conceptions of knowledge as prior to issues of method, and outline epistemological positions adequate to the knowledge produced by interviews, such as phenomenological, hermeneutical, and pragmatic philosophies, as well as narrative and discursive approaches. And we conceive of interview research as a social practice, with an emphasis on the ethical aspects of interview practice and the embeddedness of interview research in a sociopolitical context.

⊰ ONE ⊱

INTRODUCTION TO
INTERVIEW RESEARCH

———◆◆◆———

I n this chapter, we address research interviews as conversations and present different examples of interview sequences. We briefly outline a history of interviewing and depict a current interview society. We go on to outline the methodological and ethical issues in using conversations for research purposes, and conclude the chapter with an overview of the book.

CONVERSATION AS RESEARCH

The qualitative research interview attempts to understand the world from the subjects' points of view, to unfold the meaning of their experiences, to uncover their lived world prior to scientific explanations. Interview research may to some appear a simple and straightforward task. It seems quite easy to obtain a sound recorder and ask someone to talk about his or her experiences regarding some interesting topic or to encourage a person to tell his or her life story. It seems so simple to interview, but it is hard to do well. Research interviewing involves a cultivation of conversational skills that most adult human beings already possess by virtue of being able to ask questions.

There are multiple forms of conversations—in everyday life, in literature, and in the professions. Everyday conversations may range from chat and small talk to exchanges of news, disputes, or formal negotiations to deep personal interchanges. Within literature, the varieties of conversation are found in dramas, novels, and short stories, which may contain longer or shorter passages of conversations. Professional conversations include journalistic interviews, legal interrogations, academic oral examinations, philosophical dialogues, religious confessions, therapeutic sessions, and—as discussed here—qualitative research interviews. These conversational genres use different rules and techniques.

Different forms of interviews serve different purposes: Journalistic interviews are means of recording and reporting important events in society, therapeutic interviews seek to improve debilitating situations in people's lives, and research interviews have the purpose of producing knowledge. However, there are not necessarily hard-and-fast distinctions between these interview forms, for qualitative research interviews sometimes come close to journalistic interviews (and vice versa), and some qualitative researchers depict their interview practice as a therapeutic process of instigating changes in people's lives.

The research interview is based on the conversations of daily life and is a professional conversation; it is an inter-view, where knowledge is constructed in the inter-action between the interviewer and the interviewee. An interview is literally an *inter view,* an inter-change of views between two persons conversing about a theme of mutual interest. The interdependence of human interaction and knowledge production is a main theme throughout this book. In what follows, we use the term *knowledge* in a comprehensive sense, covering both everyday knowing and systematically tested knowledge.

The ambiguous drawing in Figure 1.1 was introduced by the Danish psychologist Rubin as an example of the figure/ground phenomenon in visual Gestalt perception—it can be seen alternatively as two faces or as a vase, but not as both at the same time. We use the figure to illustrate the present perspective on the interview conversation as *inter views.* We can focus on the two faces in the ambiguous figure, see them as the interviewer and the interviewee, and conceive of the interview as interaction between the two persons. Or we can focus on the vase between the two faces and see it as containing the knowledge constructed *inter* the *views* of the interviewer and the interviewee. There is an alternation between the knowers and the known, between the constructors of knowledge and the knowledge constructed. This dual aspect of the interview—the personal interrelation and the inter-view knowledge that it leads to—will run through the chapters of this book, which alternate between focusing on the personal interaction and on the knowledge constructed through that interaction.

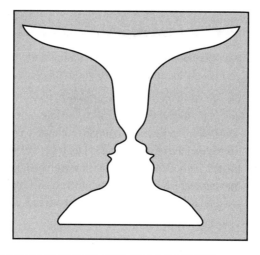

Figure 1.1   The Research Interview Seen as Inter Views

THREE INTERVIEW SEQUENCES

The use of the interview as a research method is nothing mysterious: An interview is a conversation that has a structure and a purpose. It goes beyond the spontaneous exchange of views in everyday conversations, and becomes a careful questioning and listening approach with the purpose of obtaining thoroughly tested knowledge. The research interview is not a conversation between equal partners, because the researcher defines and controls the situation. The interview researcher introduces the topic of the interview and also critically follows up on the subject's answers to his or her questions. One form of research interview—a semi-structured life world interview—will be the main focus of this book. It is defined as an interview with the purpose of obtaining descriptions of the life world of the interviewee in order to interpret the meaning of the described phenomena.

To give an initial idea of what qualitative research interviews can look like, we present three interview sequences from different research projects. The research explored Danish pupils' views on grading in high schools in Denmark, Canadian teachers' views on their work situation in a postmodern society, and the views of downtrodden youth on their living conditions in a French suburb. The passages below serve to give a first impression of qualitative research interviews.

BOX 1.1   An Interview About Grading

Interviewer:	You mentioned previously something about grades, would you please try and say more about that?
Pupil:	Grades are often unjust, because very often—very often—they are only a measure of how much you talk, and how much you agree with the teacher's opinion. For instance, I may state an opinion on the basis of a tested ideology, and which is against the teacher's ideology. The teacher will then, because it is his ideology, which he finds to be the best one, of course say that what he is saying is right and what I am saying is wrong.
Interviewer:	How should that influence the grade?
Pupil:	Well, because he would then think that I was an idiot, who comes up with the wrong answers.
Interviewer:	Is this not only your postulate?
Pupil:	No, there are lots of concrete examples.

The interview sequence in Box 1.1 is taken from a study led by the first author on the effects of grading in Danish high schools; the interview was conducted by a student taking part in the research project (Kvale, 1980); the overall design of the study is presented later (in Box 6.4). We see how the pupil, in a response to an open question from the interviewer, introduces an important dimension of his experience of grades—they are unfair—and then spontaneously provides several reasons why they are unfair. The interviewer follows up on the answers, asks for specifics, and tests the strength of the pupil's belief through counter-questions in which he doubts what the pupil tells him. This rather straightforward questioning contrasts with the reciprocity of everyday conversations. The interviewer is cast in a power position and sets the stage by determining the topic of the interchange; it is the interviewer who asks and the interviewee who answers. The researcher does not contribute with his position on the issue, nor does the pupil ask the interviewer about his view of grades.

The next sequence is from Hargreaves's interview study of the work situation of Canadian teachers and their experience of the effects of changes of school leadership in a postmodern society. One key theme that emerged was the tension between individualism and collegiality.

BOX 1.2   An Interview on Teamwork

Teacher: It's being encouraged more and more. They've been through all the schools. They want you working as a team.

Interviewer: Do you think that's good?

Teacher: So long as they allow for the creativity of the individual to modify the program. But if they want everything lock-stepped, identical—no, I think it would be disastrous, because you're going to get some people that won't think at all, that just sit back and coast on somebody else's brains, and I don't feel that's good for anybody.

Interviewer: Do you feel you're given that space at the moment?

Teacher: With [my teaching partner] I am. I know with some others here, I wouldn't [be] . . . I'd go crazy.

Interviewer: How would that be . . . ?

Teacher: Basically controlled. They would wan—first of all it would be their ideas. And I would have to fit into their teaching style, and it would have to fit into their time slot. And I don't think anybody should have to work like that.

SOURCE: *Changing Teachers, Changing Times: Teachers' Work and Culture in a Postmodern Age* (pp. 178–179), by A. Hargreaves, 1994, New York: Teachers College Press.

The teacher quoted in Box 1.2 is rather critical of the school administration's requirements of teamwork, which he regards as a control mechanism, counteracting creative teaching by the individual teachers. The interviewer does not merely register the teacher's opinions, but also asks for elaborations and receives the teacher's arguments for why he does not think that anybody should have to participate in the kind of teamwork he is subjected to. Hargreaves interpreted this and other interview sequences on teamwork as expressions of a "contrived collegiality" (see Chapter 14, in the section "Interview Analysis as Theoretical Reading").

The next sequence is from a large interview project on the conditions of downtrodden youth by the French sociologist Pierre Bourdieu and his colleagues.

BOX 1.3   An Interview With Two Young Men

-*[Pierrre Bourdieu as the interviewer] You were telling me that it wasn't much fun around here, why? What is it, your job, your leisure time?*

François: Yeah, both work and leisure. Even in this neighbourhood there is nothing much.

Ali: There's no leisure activities.

François: We have this leisure center but the neighbors complain.

Ali: They're not very nice, that's true.

- *Why do they complain, because they. . . .*

François: Because we hang around the public garden, and in the evening here is nothing in our project, we have to go in the hallways when it's too cold outside. And when there's too much noise and stuff, they call the cops.

[. . .]

- *You are not telling me the whole story . . .*

Ali: We are always getting assaulted in our project; just yesterday we got some tear gas thrown at us, really, by a guy in an apartment. A bodybuilder. A pumper.

- *Why, what were you doing, bugging him?*

François: No, when we are in the entryway he lives just above, when we are in the hall we talk, sometimes we shout.

- *But that took place during the daytime, at night?*

François: No, just in the evening.

- *Late?*

François: Late, around 10, 11 o'clock.

- *Well you know, he's got the right to snooze. The tear gas is a bit much but if you got on his nerves all night, you can see where he's coming from, right?*

> **Ali:** Yeah, but he could just come down and say. . . .
>
> *- Yes, sure, he could come down and merely say "go somewhere else" . . .*
>
> **Ali:** Instead of tear gas.
>
> SOURCE: Excerpts from Pierre Bourdieu et al. Translated by Priscilla Parkhurst Perguson and Others. "The Weight of the World" © 1999 Polity Press for translation; 1993 Editions du Seuil.

The sequence in Box 1.3 is taken from one of the many interviews reported at length by Bourdieu and his colleagues in their book on the situation of the immigrants and the poor in France (Bourdieu et al., 1999, pp. 64–65). The two young men in the interview are living in a suburban housing project in the north of France under dismal living conditions. A decade later, in late 2005, there were large uprisings among the youth in these French suburbs, protesting against their miserable situation and the harassment by the police. In this interview, Bourdieu is not a neutral questioner, but expresses his own attitudes and feelings toward the situation of the young men, and also confronts their accounts critically.

These three interviews address important issues of the subjects' life worlds, such as grades in school, changes in school leadership, and deplorable suburban living conditions. The interviewers are not merely "tape recording sociologies," to use Bourdieu's expression, but actively following up on the subjects' answers, seeking to clarify and extend the interview statements. This involves posing critical questions to the Danish pupil who believes that his teachers' grading is biased, obtaining reasons for the Canadian teacher's rejection of teamwork, and challenging the young men's presentations of themselves as innocent victims of harassment in their French suburb. We return to the social interaction and the knowledge production in these interview sequences throughout the book.

INTERVIEW RESEARCH IN HISTORY AND IN THE SOCIAL SCIENCES

Conversations are an old way of obtaining systematic knowledge. In ancient Greece Thucydides interviewed participants from the Peloponnesian Wars to write the history of the wars, and Socrates developed philosophical knowledge through dialogues with his Sophist opponents. The term *interview,* however, is of rather recent origin; it came into use in the 17th century.

BOX 1.4   Journalist Interviews

The first journalist interviews appeared in the middle of the 19th century. This form of interview has been defined by the *Oxford English Dictionary* as a "face to face meeting for the purpose of a formal conference, between a representative of the press and someone from whom he wishes to obtain statements for publication" (Murray, Bradley, Craigie, & Onions, 1961). The credit for having introduced the journalist interview has been given to Horace Greely, editor of the *New York Herald Tribune*. His interview with Brigham Young, the leader of the Mormon Church, was published in 1859 (see Silvester, 1993). Although the use of interviews in newspapers quickly caught on, they were also controversial, as the following quotes testify.

The interview is the worst feature of the new [journalism]—it is degrading to the interviewer, disgusting to the interviewee, and tiresome to the public.

—Le Figaro, 1886

Why do I refuse to be interviewed? Because it is immoral! It is a crime, just as much a crime as an offence against my person, as an assault, and just as much merits punishment. It is cowardly and vile. No respectable person would ask it, much less give it.

—Rudyard Kipling, 1892

Being interviewed does have the advantage of self-revelation. I must articulate my feelings, and I may learn something about myself. It makes me more self-aware, more aware of my own unhappiness.

—Tennessee Williams, ca. 1982

SOURCE: *The Penguin Book of Interviews: An Anthology From 1859 to the Present Day,* edited by E. Silvester, 1993, London: Penguin.

As we see in Box 1.4, the interview has not always been taken for granted as a popular form of social practice. In these statements from the early years of journalism, interviews are perceived as a somewhat "dangerous" practice that can result in immorality and unhappiness. This attitude to being interviewed contrasts rather sharply with the current age, when many people—at least in the Western world—are more than willing to be interviewed for newspapers, magazines, talk shows, and so on.

Qualitative interviews were used to varying extents in the social sciences throughout the 20th century. Although systematic literature on research interviewing is a phenomenon of the last few decades, anthropologists and sociologists have long used informal interviews to obtain knowledge from their informants. In particular, the Chicago School in sociology, which studied the urban experience in Chicago in the 1930s and 1940s, can be mentioned as an important forerunner of the later interest in qualitative research interviewing, although Chicago School researchers were rather short on methodological treatises and "just did their job" (Warren, 2002, p. 86). Survey interviews, following standard procedures with fixed wordings and sequences of questions as well as quantification of answers, have been used more frequently in the social sciences than the open qualitative interviews treated in this book. Within education and the health sciences, qualitative interviews have been a common research method for decades. Turning to our own discipline, interviews as a research method have until recently hardly been mentioned in textbooks on psychological methods, although qualitative interviews throughout the history of psychology have been a key method for producing scientific and professional knowledge. Here we briefly mention four examples of historically significant interview studies in the psychological domain.

- Freud's psychoanalytic theory was to a large extent founded on therapeutic interviews with his patients. His several hundred interviews, an hour long with each patient, were based on the patient's free associations and on the therapist's "even-hovering attention" (Freud, 1963). These qualitative interviews produced new psychological knowledge about dreams and neuroses, personality and sexuality—knowledge that after a hundred years still has a prominent position in psychological textbooks. Psychoanalysis continues to have an impact on the profession of psychotherapy, to be of interest to other disciplines and the general public, and to represent a challenge to philosophers.

- Piaget's (1930) theory of child development was based on his interviews with children in natural settings, which were often conducted in combination with simple experimental tasks. He was trained as a psychoanalyst, and what he termed his "clinical method" was inspired by the psychoanalytic interview. He let the children talk spontaneously about the weight and size of objects and using a combination of naturalistic observations, simple experiments, and interviews noticed the manner in which their thoughts unfolded.

• Experiments on the effects of changes in illumination on production at the Hawthorne Chicago plant of Western Electrical Company in the 1920s had led to unexpected results—work output and worker morale improved when the lighting of the production rooms was increased, as well as when it was decreased. These unforeseen findings were followed up in what may have been the largest interview inquiry ever conducted. More than 21,000 workers were each interviewed for over an hour and the interview transcripts analyzed qualitatively and quantitatively. The Hawthorne studies were initiated by Mayo and carried out by Roethlisberger and Dickson (1939). The researchers were inspired by therapeutic interviews and they mention the influence of Janet, Freud, Jung, and in particular Piaget, whose clinical method of interviewing children they found particularly useful. As Mayo recounted, it was necessary to train interviewers "how to listen, how to avoid interruption or the giving of advice, how generally to avoid anything that might put an end to free expression in an individual instance" (Mayo, 1933, p. 65).

• The design and advertisements of consumer products have since the 1950s been extensively investigated by individual qualitative interviews, and in recent decades by interviews in focus groups. One pioneer, Dichter, in his book *The Strategy of Desire,* reported an interview study he conducted in 1939 on consumer motivation for purchasing a car, with more than a hundred detailed conversational interviews. One main finding was how the importance of a car goes beyond its technical qualities to also encompass its "personality," which today is commonplace knowledge in marketing. Dichter described his interview technique as a "depth interview," inspired by psychoanalysis and the nondirective therapy of Carl Rogers.

These historical interview studies have made a difference to their fields, influenced the way we think about men and women and children today, and had a major impact on social practices such as therapy and techniques for controlling the behaviors of workers and consumers (Kvale, 2003). Freud and Piaget, whose main empirical evidence came from interviews, are still among the psychologists most quoted in the scientific literature, and their interpretations of their interviews with patients and children have had a major impact on how we understand personality and childhood. Thus, in *Time* magazine's selection of the 100 most influential people of the 20th century, three social scientists were among the 20 leading "scientists and thinkers": the economist Keynes and the psychologists Freud and Piaget ("Scientists and Thinkers,"

1999). The Hawthorne investigations have had a strong influence on the organization of industrial production by instigating the change from a harsh "human engineering" to a softer emotional "human relations" mode of managing workers. The marketing of consumer products today rests heavily upon qualitative interviews, in particular upon focus groups, to secure maximum prediction and control of the consumers' purchasing behaviors.

In the social sciences, qualitative interviews are now increasingly employed as a research method in their own right, with an expanding methodological literature on how to carry out interview research. Glaser and Strauss's sociological study of hospitals, reported in *The Discovery of Grounded Theory: Strategies for Qualitative Research* (1967), pioneered a qualitative research movement in the social sciences. The researchers integrated qualitative interviews into their field studies of the hospital world. Two important early books that systematically introduced research interviewing were Spradley's *The Ethnographic Interview* (1979) and Mishler's *Research Interviewing: Context and Narrative* (1986). For an overview of the scope of research interviewing, the reader is referred to Fielding's four-volume *Interviewing* (2003) and Gubrium and Holstein's *The Handbook of Interview Research* (2002), and for qualitative research more broadly to Denzin and Lincoln's *The SAGE Handbook of Qualitative Research* (2005).

Qualitative methods—ranging from participant observation to interviews to discourse analysis—have since the 1980s become key methods of social research. The rapidly growing number of books about qualitative research is one indication of this trend, thus for one leading company—Sage Publications—there was a growth in qualitative texts from 10 books in 1980–1987 to 130 books in 1995–2002 (Seale et al., 2004).

Technical, philosophical, and cultural reasons may be suggested for the growing use of qualitative research interviews. The availability of small portable tape recorders in the 1950s made the exact recording of interviews easy. In the 1980s, computer programs facilitated qualitative analyses of transcribed interviews. An opening of the social sciences to philosophy and the humanities also has taken place, drawing on phenomenology and hermeneutics as well as narrative, discursive, conversational and linguistic forms of analysis (see, e.g., Schwandt, 2001). Broad movements in philosophy influencing current social science emphasize key aspects of knowledge relevant to interview research. These aspects are the phenomenological descriptions of consciousness and of the life world, the hermeneutic interpretations of the meaning of texts, and the postmodern emphasis on the social construction of knowledge. While such fundamental philosophical positions can be at odds with one another, they have in common a rejection of a methodological positivism in the social sciences that confines scientific evidence to quantifiable facts.

Qualitative research methods in general have thus become endemic today in many disciplines, such as education, psychology, anthropology, sociology, media studies, human geography, marketing, business, and nursing science. At the backdrop to the increasing popularity of qualitative methods stands what may be called a qualitative stance. From this stance, the processes and phenomena of the world are described before theorized, understood before explained, and seen as concrete qualities before abstract quantities. The qualitative stance involves focusing on the cultural, everyday, and situated aspects of human thinking, learning, knowing, acting, and ways of understanding ourselves as persons, and it is opposed to "technified" approaches to the study of human lives.

THE INTERVIEW SOCIETY

Interviews have also become part of the common culture. In the current age, as visualized by the talk shows on TV, we live in an "interview society" (Atkinson & Silverman, 1997), where the production of the self has come in focus and the interview serves as a social technique for the public construction of the self. In Box 1.5 below, we present some impressions of the current interview society, as seen from the point of view of an interview researcher. While Atkinson and Silverman regard the interview society as one that relies "pervasively on face-to-face interviews to reveal the personal, the private self of the subject" (p. 309), we use the term somewhat more broadly to also capture the spread of interviews to a wide variety of social arenas.

BOX 1.5   A Day in the Interview Society: A Personal Account

Like most mornings, I am awakened by my clock radio. A politician is interviewed about why he has left his party. The interviewer is interested not just in his political reasons for the decision, but also in his personal motives, experiences, and hopes for the future: "Please, tell us the story about how you felt after having made the decision." Then, browsing through the main headlines of the newspaper, I notice that most articles contain statements from interviews.

I take my children to the daycare center, and their teacher greets me with a probing interview about the kids and their opportunities for further development. Having read some literature on current educational practices, I recognize her conversational technique as a version of appreciative inquiry, a style of interviewing that focuses on positive experiences and narratives.

When I get to work, I have scheduled a meeting with a journalist, who wants to interview me about a book that I have contributed to. I try to think of something interesting to say—in today's world of media, with their cacophony of competing voices, journalists are always looking for a new angle when they do interviews. I then engage in a session of student supervision, which involves a subtle interviewing technique of its own. The goal is to ask questions that will help the student progress, rather than posing examining or confusing questions.

After work, I ask my wife about how her day has been, and she tells me about her experiences. Being a schoolteacher, she reports a narrative of how she has struggled with ways of getting the pupils to say something in class. How can she improve her questioning techniques? When she then asks me about my experiences, the phone interrupts us—a market researcher wants to interview me about my consumption habits concerning breakfast products. I decide to be a reluctant respondent and quickly end the conversation.

Watching the news on TV, I am confronted with interviews with business leaders, politicians, and also ordinary men and women, who seem more than happy to express their opinions. A witness to a traffic accident is interviewed, and someone else is interviewed about how he feels about a certain politician's trustworthiness. After the news, I watch one of the many confessional talk shows that run on TV, where the host manages to do three life story interviews in half an hour.

I find that I am unable to fall asleep—perhaps due to the conversational bombardment I have experienced—and I put on one of my favorite Woody Allen DVDs. I come to the conclusion that Allen's movies are a perfect representation of the interview society. In almost every scene, people are constantly talking, interviewing each other and even themselves, while walking, eating, partying, in therapy, and having sex. Allen's characters live in a truly conversational reality, where experiences, desires, and doubts are relentlessly shared, and the self is expressed and constructed in and through speaking. I finally manage to fall asleep with the 6-pound *SAGE Handbook of Qualitative Research* in my hands.

Box 1.5 describes a day in the interview society of the second author. A variety of different interview forms were encountered, some were informal and were simply conversations as goals in themselves (e.g., conversations at the dinner table), and others were structured as professional conversations with specific goals (to inform, supervise, or entertain an audience, for example). Whereas journalistic interviews were regarded as somewhat dangerous practices in the 19th century, in the 20th century they became simply taken for granted as a standard form of human relations.

Atkinson and Silverman (1997) attributed the contemporary prevalence of interviews to a spirit of the age; we would like to add that there is a more material basis of the rise of interviewing—the experiential economy of a consumer society. With knowledge of consumers' experiences and lifestyles so essential to the Western economy, qualitative interviewing for consumer experiences, in particular in the form of focus groups, has become a key approach to predicting and controlling consumer behavior.

METHODOLOGICAL AND ETHICAL ISSUES IN RESEARCH INTERVIEWING

We now turn from discussing the pervasive role of interviews in the broader social scene to depicting some of the methodological and ethical concerns that we address throughout this book. While one form of research interview—a semi-structured life world interview, in part inspired by phenomenology—will be our focus, other ways of conducting and analyzing interviews will also be treated. Whereas phenomenologists are typically interested in charting how human subjects *experience* life world phenomena, hermeneutical scholars address the *interpretation* of meaning, and discourse analysts focus on how language and discursive practices *construct* the social worlds in which human beings live. This book does not aim to settle the questions concerning these epistemological and ontological differences between different philosophies of qualitative inquiry, but instead has a pragmatic approach. In this approach, reflections on how to conduct and analyze interviews are based on what the researcher is interested in knowing about: Is it primarily experiences of concrete episodes, the meanings of specific phenomena, comprehension of specific concepts, processes of discursive construction, or something different? Our aim is not to force certain philosophical preconceptions onto our readers, but to assist you in making informed choices about what to do when conducting interview research, reflected choices that we hope will allow you to engage more deeply with the kind of knowledge you will be producing in your research interviews.

The closeness of the research interview to everyday conversation may imply a certain simplicity, but this simplicity is illusory. Nevertheless, it has contributed to the popularity of research interviewing—it is too easy to start interviewing without any preceding preparation or reflection. A novice researcher may have a good idea, grab a sound recorder, go out and find some subjects, and start questioning them. The recorded interviews are transcribed and then—during the analysis of the many pages of transcripts—a multitude of problems about the purpose and content of the interviews surfaces. There is little likelihood that such spontaneous interview studies will lead to worthwhile information; rather than producing new substantial knowledge about a topic, such interviews may be reproducing common opinions and prejudices. That being said, interviewing can be an exciting way of doing strong and valuable research. The unfolding of stories and new insights can be rewarding for both parties in the interview interaction. Reading the transcribed interviews may inspire the researcher to new interpretations of well-known phenomena, and the interview reports can contribute substantial new knowledge to a field.

A novice researcher who is methodologically oriented may have a host of questions about the technical and conceptual issues in an interview project. For example: How do I begin an interview project? How many subjects will I need? How can I avoid influencing the subjects with leading questions? Can the interviews be harmful to the subjects? Is transcription of the interviews necessary? How do I analyze the interviews? Will my interpretations only be subjective? Can I be sure that I get to know what the subjects really mean? How do I report my extensive interview texts?

If corresponding questions were raised about, for example, a questionnaire survey, several of them would be fairly easy to answer by consulting authoritative textbooks on standard techniques and rules of survey research. As this book makes clear, the situation is quite the contrary for the craft of qualitative interview research, for which there are few standard rules or common methodological conventions. Interview research is a craft that, if well carried out, can become an art. The varieties of research interviews approach the spectrum of human conversations. The forms of interview analysis can differ as widely as ways of reading a text. The qualitative interview is some-times called an *unstructured* or a *nonstandardized* interview. Because there are few prestructured or standardized procedures for conducting these forms of interviews, many of the methodical decisions have to be made on the spot, during the interview. This requires a high level of skill on behalf of the inter-viewer, who needs to be knowledgeable about the interview topic and to be familiar with the methodological options available, as well as to have an understanding of the conceptual issues of producing knowledge through conversation.

In this book we will attempt to steer between the free spontaneity of a no-method approach to interviewing and the rigid structures of an all-method approach by focusing on the expertise, skills, and craftsmanship of the inter-view researcher. Some of the decisions that will have to be made on the way through the stages of an interview inquiry and the methodological options available are outlined, and the specific modes of questioning are discussed, as well as the multiple options for analyzing interviews. If one is looking for a cookbook approach to the practice of qualitative research interviewing, how-ever, reading this book may be frustrating; in lieu of standard procedures and fixed rules, the answers to questions such as those posed above will most often be prefaced by "It depends," as the answers depend upon the specific purpose and topic of an investigation.

Ethical issues permeate interview research. The knowledge produced by such research depends on the social relationship of interviewer and intervie-wee, which rests on the interviewer's ability to create a stage where the sub-ject is free and safe to talk of private events recorded for later public use. This again requires a delicate balance between the interviewer's concern for pursu-ing interesting knowledge and ethical respect for the integrity of the interview subject. The tension between the pursuit of knowledge and ethics in research interviewing is well expressed in Sennett's 2004 book *Respect*:

> In-depth interviewing is a distinctive, often frustrating craft. Unlike a pollster asking questions, the in-depth interviewer wants to probe the responses people give. To probe, the interviewer cannot be stonily impersonal; he or she has to give something of himself or herself in order to merit an open response. Yet the conversation lists in one direction; the point is not to talk the way friends do. The interviewer all too frequently finds that he or she has offended subjects, transgressing a line over which only friends or intimates can cross. The craft consists in calibrating social distances without making the subject feel like an insect under the microscope. (pp. 37–38)

OVERVIEW OF THE BOOK

Throughout this book we will follow three main lines to guide the learning of interviewing for research purposes: We will approach interviewing as a craft, as a knowledge-producing activity, and as a social practice.

Interviewing as a Craft

Interviewing rests on the practical skills and the personal judgments of the interviewer; it does not follow explicit steps of rule-governed methods. The skills of interviewing are learned by practicing interviewing, and the quality of interviewing is judged by the strength and value of the knowledge produced. The conception of interviewing as a craft, to be learned through practice, contrasts with a methodological positivism in the social sciences, with its conception of research as following the rules and predetermined steps of specific methods. In our pragmatic craft approach, we do not attempt to derive rules of an interview method from some normative theory of science, but attempt to learn from how competent interview researchers work. This book seeks to promote learning through exemplary cases; it presents and discusses examples from interview studies, some of which have made significant differences to their respective fields. Rules of thumb for interviewing, derived from interview practice, are presented, and for the interviewer craftsman (the term *craftsman* as it is used in this book applies to interviewers of both genders) we provide toolboxes with a variety of techniques, in particular for the key stages of conducting and analyzing interviews.

The skills of interviewing are learned through the practice of interviewing, and there is a paradox in presenting a textbook for the learning of a skill, a paradox enhanced by addressing in written form the learning of an oral skill. We return to this paradox in the final chapter, but throughout this book we engage with the paradox by giving suggestions for learning interviewing the way a craft is learned. The book depicts the journey through the practical stages of an interview project, providing the necessary road directions and equipment. The chapters examine the complex skills of the interview craft, breaking them down into discrete steps, giving examples, and pointing out the practical, conceptual, and ethical issues involved.

Interviewing as a Social Production of Knowledge

Interviewing is an active process where interviewer and interviewee through their relationship produce knowledge. Interview knowledge is produced in a conversational relation; it is contextual, linguistic, narrative, and pragmatic. The conception of interview knowledge presented here contrasts with a methodological positivism conception of knowledge as given facts to be quantified. This book presents philosophies congenial to the knowledge produced in interview research, such as phenomenology, hermeneutics, pragmatism, and postmodern thought. We show how different conceptions of interview knowledge lead to different forms of interviewing and of analyses of interviews. Some see the practice of qualitative research interviewing as involving an unearthing of preexisting meaning nuggets from the depths of the respondent, while others argue that it should be an unbound and creative process where the researcher is free to construct appealing stories. Rather than locating the meanings and narratives to be known solely in the subjects or the researchers, we argue in this book that the process of knowing through conversations is intersubjective and social, involving interviewer and interviewee as co-constructors of knowledge.

Interviewing as a Social Practice

Interviewing is a new practice of the last few centuries; today it has become a pervasive social practice in what has been called the interview society. Interviewing as a mode of inquiry is embedded in a historical and social context. The interaction of interviewer and interviewee is laden with ethical issues, and publishing interview research entails broader sociopolitical concerns. We address ethical issues of the specific interview practices as well as the social effects of interview research, and also take issue with a belief in interviewing as a particularly ethical form of research. Our conception of interviewing as social practice contrasts with the idealism in many textbooks, which present interviewing predominantly within the context of ideas and as a pure and authentic interaction within a human relationship. This book addresses the power asymmetry of the interview situation and also includes the broader social influences upon, and social consequences of, interview research. To these social influences and consequences belong the current impact of ethical review boards and evidence-based practice upon research interviewing, the consequences of interviews in co-shaping our conceptions of human beings, and in providing knowledge for human management and manipulation, as well as contributing to public enlightenment.

Part I: Conceptualizing the Research Interview

In this conceptual part of the book, we address principal issues concerning the use of conversations for research purposes. While it may appear as a truism that theory and practice should be related in the production of interview knowledge, the issues become more complex when we turn to how conceptions of knowledge and practice are related.

We display the richness and varieties of conversations and give examples of a research interview, a philosophical dialogue, and a therapeutic interview. We depict philosophical approaches that are open to qualitative interviewing, such as postmodern thought, phenomenological, hermeneutical, and pragmatic philosophy. We address ethical issues in research interviewing, arguing that ethical research behavior involves more than rule following and adherence to ethical codes. Ethics is basic to an interview inquiry; it goes beyond ethical rules to encompass the broader fields of ethical and sociopolitical uncertainties in social research. In situations of conflict, decisions about which rules to follow will to a large extent depend upon the researcher's experience and personal judgment. As a bridge to Part II, which focuses on the practice of interviewing, we conclude the first part with a discussion of viewing research interviewing as a method or a craft. A conception of research interviewing as a rule-governed method will lead to different interview practices than an understanding of research interviewing as a craft, where the quality of the interview knowledge rests upon the skills and the personal judgment of the interviewer craftsman.

NB: Readers who want to learn interviewing by doing interviews should discontinue reading the book now and jump to the appendix, Learning Tasks, where we suggest exercises for those interested in learning interviewing in ways that approximate the learning of a craft. After spending a few weeks with these tasks, preferably in the company of co-learners, return to reading the book and you will discover that by practicing interviewing, you may already have started upon some of the theoretical reflections put forth in the conceptual chapters of Part I and have experienced a good deal of what is said about conducting interviews in the practical chapters of Part II.

Part II: Seven Stages of Research Interviewing

Part II treats in detail the practical steps of interviewing. The chapters follow seven stages of an interview investigation: (1) thematizing an interview project, (2) designing, (3) interviewing, (4) transcribing, (5) analyzing, (6) verifying, and (7) reporting. The importance of conceptualizing an interview topic in advance of interviewing, as well as planning an entire interview project through seven stages before starting to interview, is pointed out. The chapters provide the interviewer craftsman with toolboxes for the different stages of his or her journey. Although we go into detail with the common life world interview, other forms of interviewing also are addressed, such as narrative and discursive interviews, as well as more confrontational interviews. When it comes to the analyses of the interviews, different forms of categorization, condensation, and interpretation of meanings are depicted, as well as linguistic analyses in the form of conversational, narrative, and discursive approaches.

Concluding Perspectives

In the concluding chapter, Chapter 17, we summarize the three main lines that are followed through the book: interviewing as a craft, as a knowledge producing activity, and as a social practice. We suggest potentials for developing the quality of interview research and conclude by emphasizing a pragmatic validation of interview research through producing knowledge worth knowing—knowledge that makes a difference to a discipline and those who depend on it.

CONCEPTUALIZING
THE RESEARCH INTERVIEW

———◆———

I n this first part of the book, we discuss principal issues of research interviewing, particularly those related to epistemology and ethics. We address
the qualitative research interview as a specific form of conversation; in
Chapter 2, we exemplify and outline the mode of understanding of a qualitative research interview and relate it to a philosophical dialogue and a therapeutic interview. We then treat epistemological questions concerning the
production of interview knowledge in Chapter 3. In Chapter 4, we address the
ethical dimension of the social practice of interviewing, and in Chapter 5 we
argue that knowledge production in interview research comes closer to a craft
than a rule-following method. We point out some implications of these conceptual issues for practicing interview research, implications we treat in more
detail in relation to the seven stages of an interview investigation in Part II,
which covers the practice of qualitative research interviewing.

A reader who is impatient to learn the practical skills of the interview craft
may ask why he or she should bother with such complex and subtle issues and
not just go straight to the practice of interviewing. One answer is that an able
craftsman needs to be familiar with the materials he or she is working with and
also the product that is the goal. The research interviewer works with language

and knowledge, and the product arrived at is likewise knowledge in a linguistic form. The able interviewer is familiar with the nuances and problems of the material she works with and with the value and strength of the product she delivers.

Readers who are unfamiliar with social science research and philosophy may go directly to Part II, where the practice of research interviewing is presented in seven stages, and then return to the conceptual issues after becoming more familiar with interview practice.

TWO

RESEARCH INTERVIEWS, PHILOSOPHICAL DIALOGUES, AND THERAPEUTIC INTERVIEWS

———◆———

This chapter begins with a research interview about learning, based on a phenomenological approach (in later chapters, we present alternatives to phenomenological life world interviewing). After a brief outline of phenomenology follows a depiction inspired by phenomenology of the mode of understanding in a qualitative research interview. In contrast to the common emphasis on empathy and equality in qualitative interviewing, we point out the power asymmetry of a research interview. We then go on to highlight the specific nature of the research interview by comparing and contrasting it with two other forms of interviews—the philosophical dialogue and the therapeutic interview. We compare and contrast the modes of interaction and understanding in the research interview with the logical/cognitive mode of philosophical dialogues and the emotional/personal mode of therapeutic interviews. We present a philosophical dialogue by Socrates, then discuss the logic of this Socratic form of interview inquiry and show its relationship to current research interviewing. Finally, we present a therapeutic interview, outline one mode of understanding in therapeutic interviews, and mention implications for the history of research interviewing.

A QUALITATIVE RESEARCH INTERVIEW ON LEARNING

The purpose of the qualitative research interview discussed here is to understand themes of the lived daily world from the subjects' own perspectives. The structure comes close to an everyday conversation, but as a professional interview, it involves a specific approach and technique of questioning. The following interview passage is taken from the article "An Application of Phenomenological Method in Psychology" by Giorgi (1975). The research question guiding the interview was: What constitutes learning in the everyday world? The first half of the interview, conducted by a student, is reproduced here.

R (Researcher): Could you describe in as much detail as possible a situation in which learning occurred for you?

S (Subject): (E. W., 24-year-old female, housewife and educational researcher): The first thing that comes to mind is what I learned about interior decorating from Myrtis. She was telling me about the way you see things. Her view of looking at different rooms has been altered. She told me that when you come into a room you don't usually notice how many vertical and horizontal lines there are, at least consciously, you don't notice. And yet, if you were to take someone who knows what's going on in the field of interior decorating, they would intuitively feel if there were the right number of vertical and horizontal lines. So, I went home, and I started looking at the lines in our living room, and I counted the number of horizontal and vertical lines, many of which I had never realized were lines before. A beam . . . I had never really thought of that as vertical before, just as a protrusion from the wall. (Laughs) I found out what was wrong with our living room design: many, too many, horizontal lines and not enough vertical. So I started trying to move things around and change the way it looked. I did this by moving several pieces of furniture and taking out several knick-knacks, de-emphasizing certain lines, and . . . it really looked differently to me. It's interesting because my husband came home several hours later and

I said, "Look at the living room; it's all different." Not knowing this, that I had picked up, he didn't look at it in the same way I did. He saw things were different, he saw things were moved, but he wasn't able to verbalize that there was a de-emphasis on the horizontal lines and more of an emphasis on the vertical. So I felt I had learned something.

R: What part of that experience would you consider learning?

S: The knowledge part that a room is made up of horizontal and vertical lines. The application of that to another room; applying it to something that had been bothering me for quite a long time and I could never put my finger on it. I think the actual learning was what was horizontal and vertical about a room. The learning that was left with me was a way of looking at rooms.

R: Are you saying then that the learning was what you learned from Myrtis, what you learned when you tried to apply . . . ?

S: Since I did apply it, I feel that I learned when I did apply it. I would have *thought* that I learned it only by having that knowledge, *but* having gone through the act of application, I really don't feel I would have learned it. I could honestly say, I had learned it at that time. (pp. 84–86)

This interview investigated what constitutes learning for a woman in her everyday world. It began with an open request to describe a situation where learning occurred. The woman herself chose the learning situation she would talk about—interior decorating; she described this freely and extensively in her own words. The answer spontaneously took the form of a story, a narrative of one learning episode. The interviewer's first question introduced learning as the theme of the interview; her remaining questions departed from the subject's answers in order to keep learning in focus and to ask for clarification of the different aspects of the subject's learning story.

This interview gives a good picture of a semi-structured research interview focusing on the subject's experience of a theme. The interviewer's questions aimed at a cognitive clarification of the subject's experience of learning, which will be further analyzed in Chapter 12. The mode of interviewing was inspired by phenomenological philosophy, to which we now turn.

PHENOMENOLOGY AND THE MODE OF UNDERSTANDING IN A QUALITATIVE RESEARCH INTERVIEW

Phenomenology was founded as a philosophy by Edmund Husserl around 1900 and further developed as existential philosophy by Martin Heidegger, and then in an existential and dialectical direction by Jean-Paul Sartre and Maurice Merleau-Ponty. The subject matter of phenomenology began with consciousness and experience, and then was expanded by Husserl and also Heidegger to include the human life world, and by Merleau-Ponty and Sartre to take account of the body and human action in historical contexts.

A phenomenological approach has, in a general nonphilosophical sense, been prevalent in qualitative research. In sociology, phenomenology was mediated by the Husserlian-based phenomenology of the social world by Alfred Schutz, and further by Berger and Luckmann in *The Social Construction of Reality* (1966) and by Garfinkel (1967) in his ethnomethodological studies of the practical production of social order. Generally, in qualitative inquiry, *phenomenology* is a term that points to an interest in understanding social phenomena from the actors' own perspectives and describing the world as experienced by the subjects, with the assumption that the important reality is what people perceive it to be. The open phenomenological approach to the meanings of phenomena in the everyday world will be taken up again when we address how to analyze interviews (in Chapter 12 on meaning condensation).

In focusing the interview on the experienced meanings of the subjects' life world, phenomenology has been relevant for clarifying the mode of understanding in a qualitative research interview. The implications of phenomenological philosophy for qualitative research were developed in a series of studies at Duquesne University. Starting with van Kaam's (1959) study of "the experience of really being understood," the method was further applied, systematized, and reflected upon by the phenomenological psychologist Giorgi and his colleagues, among others (see Fischer & Wertz, 1979; Giorgi, 1970; Giorgi & Giorgi, 2003). According to Giorgi, "Phenomenology is the study of the structure, and the variations of structure, of the consciousness to which any thing, event, or person appears" (Giorgi, 1975, p. 83).

BOX 2.1   Phenomenological Method

According to Merleau-Ponty (1962), what matters is to describe the given as precisely and completely as possible; to describe rather than to explain or analyze. It is not possible to give precise instructions for an *open description,* and Spiegelberg (1960) illustrates the method by using metaphors; for example, "to the matters themselves," "seeing and listening," "keeping the eyes open," "not think, but see."

In phenomenological philosophy, objectivity is an expression of fidelity to the phenomena investigated. The goal is to arrive at an *investigation of essences* by shifting from describing separate phenomena to searching for their common essence. Husserl termed one method of investigating essences a "free variation in fantasy." This means varying a given phenomenon freely in its possible forms, and that, which remains constant through the different variations, is the essence of the phenomenon.

A phenomenological *reduction* calls for a suspension of judgment as to the existence or nonexistence of the content of an experience. The reduction can be pictured as a "bracketing," an attempt to place the common sense and scientific foreknowledge about the phenomena within parentheses in order to arrive at an unprejudiced description of the essence of the phenomena.

In Box 2.1 we have, based on Spiegelberg (1960), outlined a phenomenological method that includes description, investigation of essences, and phenomenological reduction. We shall now depict more specifically the mode of understanding in a qualitative research interview from a perspective inspired by phenomenology.

A semi-structured life world interview attempts to understand themes of the lived everyday world from the subjects' own perspectives. This kind of interview seeks to obtain descriptions of the interviewees' lived world with respect to interpretation of the meaning of the described phenomena. It comes close to an everyday conversation, but as a professional interview it has a purpose and involves a specific approach and technique; it is semi-structured—it is neither an open everyday conversation nor a closed questionnaire. It is conducted according to an interview guide that focuses on certain themes and that may include suggested questions. The interview is usually transcribed, and the written text and sound recording together constitute the materials for the subsequent analysis of meaning.

BOX 2.2   Twelve Aspects of Qualitative Research Interviews

Life World. The topic of qualitative interviews is the everyday lived world of the interviewee and his or her relation to it.

Meaning. The interview seeks to interpret the meaning of central themes in the life world of the subject. The interviewer registers and interprets the meaning of what is said as well as how it is said.

Qualitative. The interview seeks qualitative knowledge expressed in normal language, it does not aim at quantification.

Descriptive. The interview attempts to obtain open nuanced descriptions of different aspects of the subjects' life worlds.

Specificity. Descriptions of specific situations and action sequences are elicited, not general opinions.

Deliberate Naiveté. The interviewer exhibits openness to new and unexpected phenomena, rather than having readymade categories and schemes of interpretation.

Focused. The interview is focused on particular themes; it is neither strictly structured with standardized questions, nor entirely "nondirective."

Ambiguity. Interviewee statements can sometimes be ambiguous, reflecting contradictions in the world the subject lives in.

Change. The process of being interviewed may produce new insights and awareness, and the subject may in the course of the interview come to change his or her descriptions and meanings about a theme.

Sensitivity. Different interviewers can produce different statements on the same themes, depending on their sensitivity to and knowledge of the interview topic.

Interpersonal Situation. The knowledge obtained is produced through the interpersonal interaction in the interview.

Positive Experience. A well carried out research interview can be a rare and enriching experience for the interviewee, who may obtain new insights into his or her life situation.

Box 2.2 gives an outline of twelve aspects of a qualitative interview from a phenomenological perspective. It will be treated in more detail below and in Chapter 7, whereas other forms of interviewing are addressed in Chapter 8. In what follows, we elaborate on the twelve aspects of qualitative interviewing from a phenomenological standpoint.

- *Life World.* The topic of qualitative research interviews is the interviewee's lived everyday world. The qualitative research interview has a unique potential for obtaining access to and describing the lived everyday world. The attempt to obtain unprejudiced descriptions entails a rehabilitation of the *Lebenswelt*—the life world—in relation to the world of science. The life world is the world as it is encountered in everyday life and given in direct and immediate experience, independent of and prior to explanations. The qualitative interview may be seen as one realization of Merleau-Ponty's (1962) program for a phenomenological science starting from the primary experience of the world:

> All my knowledge of the world, even my scientific knowledge, is gained from my own particular point of view, or from some experience of the world without which the symbols of science would be meaningless. The whole universe of science is built upon the world as directly experienced, and if we want to subject science itself to rigorous scrutiny and arrive at a precise assessment of its meaning and scope, we must begin by re-awakening the basic experiences of the world of which science is the second order expression. (p. viii)

The geographer's map is thus an abstraction of the countryside where we first learned what a forest, a mountain, or a river was. In this phenomenological approach, the qualitative studies of subjects' experiences of their world are basic to the more abstract scientific theories of the social world; interviews are in this sense not merely a few entertaining curiosities added to some basic scientific quantitative facts obtained by experiments and questionnaires. The qualitative interview is a research method that gives a privileged access to people's basic experience of the lived world.

- *Meaning.* The interview seeks to understand the meaning of central themes of the subjects' lived world. The interviewer registers and interprets the meanings of what is said as well as how it is said; he or she should be knowledgeable about the interview topic, be observant of—and able to interpret—vocalization, facial expressions, and other bodily gestures. An everyday conversation often takes place on a factual level. A pupil may state: "I am not as stupid as my grades at the examinations showed, but I have bad study habits." Common reactions could then concern matters of fact: "What grades did you get?" or "What are your study habits?"—questions that also may yield important information. A meaning-oriented reply would, in contrast, be something like, "You feel that the grades are not an adequate measure of your competence?"

A qualitative research interview seeks to cover both a factual and a meaning level, although it is usually more difficult to interview on a meaning level. It is necessary to listen to the explicit descriptions and to the meanings expressed, as well to what is said "between the lines." The interviewer can seek to formulate the implicit message, "send it back" to the subject, and may obtain an immediate confirmation or disconfirmation of the interpretation of what the interviewee is saying.

- *Qualitative.* The qualitative interview seeks qualitative knowledge as expressed in normal language; it does not aim at quantification. The interview aims at nuanced accounts of different aspects of the interviewee's life world; it works with words and not with numbers. The precision in description and stringency in meaning interpretation in qualitative interviews correspond to exactness in quantitative measurements.

- *Descriptive.* The qualitative interviewer encourages the subjects to describe as precisely as possible what they experience and feel, and how they act. The focus is on nuanced descriptions that depict the qualitative diversity, the many differences and varieties of a phenomenon, rather than on ending up with fixed categorizations. The question of why the subjects experience and act as they do is primarily a task for the researcher to evaluate.

- *Specificity.* Descriptions of specific situations and actions are elicited, not general opinions. On the basis of comprehensive accounts of specific situations and events, the interviewer will be able to arrive at meanings on a concrete level, instead of general opinions obtained by questions such as "What is your opinion of grading?" Still, it should be recognized that this type of general opinion question might yield information that is of interest in itself.

- *Deliberate naiveté.* The interviewer exhibits openness to new and unexpected phenomena, rather than having readymade categories and schemes of interpretation. The qualitative interview attempts to obtain descriptions that are as inclusive and presuppositionless as possible of important themes of the interviewee's life world. Rather than the interviewer posing preformulated questions with respect to prepared categories for analysis, the deliberate naiveté and a bracketing of presuppositions implies openness to new and unexpected phenomena. The interviewer should be curious, sensitive to what is said—as well as to what is not said—and critical of his or her own presuppositions and hypotheses during the interview. Thus, presuppositionlessness implies a critical awareness of the interviewer's own presuppositions.

• *Focused.* The interview is focused on particular themes; it is neither strictly structured with standard questions, nor entirely "nondirective." Through open questions the interview focuses on the topic of research. It is then up to the subject to bring forth the dimensions he or she finds important in the theme of inquiry. The interviewer leads the subject toward certain themes, but not to specific opinions about these themes.

• *Ambiguity.* The interviewee's answers are sometimes ambiguous. One statement can imply several possibilities of interpretation, and the subject may also give apparently contradictory statements during an interview. The aim of the qualitative research interview is not to end up with unequivocal and quantifiable meanings on the themes in focus. The task of the interviewer is to clarify, as far as possible, whether the ambiguities and contradictory statements are due to a failure of communication in the interview situation, or whether they reflect genuine inconsistencies, ambivalence, and contradictions in an interviewee's life situation. The contradictions of interviewees need not merely be due to faulty communication in the interview, nor to the interviewee's personality, but may be adequate reflections of objective contradictions in the world in which they live.

• *Change.* In the course of an interview, the subjects can change their descriptions of, and attitudes toward, a theme. The subjects may themselves discover new aspects of the themes they are describing, and suddenly see relations that they had not been aware of earlier. The questioning can thus instigate processes of reflection where the meanings of themes described by the subjects are no longer the same after the interview. An interview may be a learning process for the interviewee, as well as for the interviewer.

• *Sensitivity.* Different interviewers, using the same interview guide, may produce different statements on the same themes, due to varying levels of sensitivity toward, and knowledge about, the topic of the interview. Thus an interviewer who has no ear for music may have difficulties obtaining nuanced descriptions of musical experiences from his or her interviewees, in particular if the interviewer is trying to probe more intensively into the meaning of the music. If a common methodological requirement of obtaining intersubjectively reproducible data were to be followed here, the interview form might have to be standardized in a way that would restrict the understanding of musical experiences to more superficial aspects understandable to the average person. The requirement of sensitivity to, and a foreknowledge about, the topic of the interview contrasts with the presuppositionless attitude advocated above. The tension between these two aspects may be expressed in the requirement for a qualified naiveté on the part of the interviewer.

• *Interpersonal Situation.* The research interview is an inter-view where knowledge is constructed in the inter-action between two people. The interviewer and the subject act in relation to each other and reciprocally influence each other. The interaction may also be anxiety provoking and evoke defense mechanisms in the interviewee as well as in the interviewer. The interviewer should be aware of potential ethical transgressions of the subject's personal boundaries and be able to address the interpersonal dynamics within an interview. The knowledge produced in a research interview is constituted by the interaction itself, in the specific situation created between an interviewer and an interviewee. With another interviewer, a different interaction may be created and a different knowledge produced.

• *Positive Experience.* A well-conducted research interview may be a rare and enriching experience for the subject, who may obtain new insights into his or her life situation. It is probably not a very common experience in everyday life that another person—for an hour or more—shows an interest in, is sensitive toward, and seeks to understand as well as possible one's own experiences and views on a topic. In practice, it may sometimes be difficult to terminate a qualitative interview, as the subject may want to continue the conversation and explore further the insights into his or her life world brought about by the interview.

We have here attempted, inspired by phenomenology, to depict the mode of understanding in a semi-structured and empathetic life world interview, which was exemplified by the phenomenological interview on learning in everyday life.

POWER ASYMMETRY IN QUALITATIVE RESEARCH INTERVIEWS

Taking into account the mutual understanding and the personal interview inter-action described in the twelve aspects above, we should not regard a research interview as a completely open and free dialogue between egalitarian partners. The empathetic form of phenomenological life world interviewing appears har-monious, and issues of power have been little addressed in relation to these and other forms of qualitative research interviews. The research interview is, how-ever, a specific professional conversation with a clear power asymmetry between the researcher and the subject. In order to correct the potential misunderstanding of research interviews as a dominance-free zone of consensus and empathy, we shall point out some power asymmetries in qualitative research interviews.

BOX 2.3   Power Asymmetry in Qualitative Research Interviews

The interview entails an asymmetrical power relation. The research inter-view is not an open everyday conversation between equal partners. The interviewer has scientific competence, he or she initiates and defines the interview situation, determines the interview topic, poses questions and decides which answers to follow up, and also terminates the conversation.

The interview is a one-way dialogue. An interview is a one-directional questioning—the role of the interviewer is to ask, and the role of the inter-viewee is to answer.

The interview is an instrumental dialogue. In the research interview an instrumentalization of the conversation takes place. A good conversation is no longer a goal in itself, but a means for providing the researcher with descriptions, narratives, texts—to interpret and report according to his or her research interests.

The interview may be a manipulative dialogue. A research interview may follow a more or less hidden agenda. The interviewer may want to obtain information without the interviewee knowing what the interviewer is after, attempting to "by indirections find directions out."

(Continued)

(Continued)

The interviewer has a monopoly of interpretation. The researcher usually has a monopoly of interpretation over the subject's statements. As the "big interpreter," the researcher maintains an exclusive privilege to interpret and report what the interviewee really meant.

COUNTER-CONTROL. In reaction to the dominance of the interviewer, some subjects will withhold information, or talk around the subject matter, and some may start to question the researcher and also protest at his or her questions and interpretations, or, in rare cases, withdraw from the interview.

EXCEPTIONS. Some interviewers attempt to reduce the power asymmetry of the interview situation by collaborative interviewing where the researcher and subject approach equality in questioning, interpreting, and reporting.

The asymmetry of the power relation in the research interviewer outlined in Box 2.3 is easily overlooked if we only focus on the open mode of understanding and the close personal interaction of the interview. There does not need to be any intentional exertion of power by the interviewer; the description concerns the structural positions in the interview, whereby for example subjects may, more or less deliberately, express what they believe the interviewer authority wants to hear. If power is inherent in human conversations and relations, the point is not that power should necessarily be eliminated from research interviews, but rather that interviewers ought to reflect on the role of power in the production of interview knowledge. Acknowledging the power relations in qualitative research interviews raises both epistemological issues about the implications for the knowledge produced and ethical issues about the implications for how to deal responsibly with power asymmetries, and we return to these questions in the two following chapters on epistemology and ethics.

PHILOSOPHICAL DIALOGUES AND RESEARCH INTERVIEWS

In order to highlight the mode of understanding, and the specific interaction in the research interview, we shall now compare it to another form of conversation—the cognitive logical argumentation of a philosophical dialogue by Socrates.

BOX 2.4   A Philosophical Dialogue on Love

"And quite properly, my friend," said Socrates; "then, such being the case, must not Love be only love of beauty, and not of ugliness?" He assented.

"Well then, we have agreed that he loves what he lacks and has not?"

"Yes," he replied.

"And what Love lacks and has not is beauty?"

"That needs must be," he said.

"Well now, will you say that what lacks beauty, and in no wise possesses it, is beautiful?"

"Surely not."

"So can you still allow Love to be beautiful, if this is the case?"

Whereupon Agathon said, "I greatly fear, Socrates, I know nothing of what I was talking about."

SOURCE: *V. Lysis, Symposion, Gorgias* (p. 167), by Plato (translated by W. R. M. Lamb), 1953, Cambridge, MA: Harvard University Press.

Plato's *Symposion* is a philosophical dialogue in a dramatic form. The partners in the dialogue are formally on an equal level; there is a reciprocal questioning of the true nature of the knowledge under debate, as well as of the logic of the participants' questions and answers. In *Symposion,* Socrates takes Agathon's speech on love as his point of departure. He repeats its main points in a condensed form, interprets what Agathon has said, and asks for his opponent's confirmations or disconfirmations of the interpretations. Socrates started out by appearing naive and innocent, then praised Agathon's views on Eros, after which he followed up by uncovering one contradiction after another in Agathon's position. This philosophical dialogue is a harsh form of interaction that seeks true knowledge through the unrelenting rigor of a discursive argumentation. Socrates compared himself to a legal interrogator and his opponents likened him to an electric eel. In Chapter 4, on ethics, we shall see another example of Socrates examining his opponents.

Dinkins (2005, p. 124) has outlined the general principles of Socratic interviewing, which she refers to as "Socratic-hermeneutic interpre-viewing." Socrates' "method" in the dialogues is not a method in the conventional sense of following a fixed procedure toward a goal, but rather an examining of a person by considering his or her statements normatively. The Socratic conversation is a fundamental mode of understanding, rather than a method in any mechanical sense. In Dinkins's rendition, Socrates' examining proceeds as follows:

1. Socrates encounters someone who takes an action or makes a statement into which Socrates wishes to inquire.

2. Socrates asks the person for a definition of the relevant central concept, which is then offered.

3. Together, Socrates and the respondent (or "co-inquirer" to use Dinkins's term) deduce some consequences of the definition.

4. Socrates points out a possible conflict between the deduced consequences and another belief held by the respondent. The respondent is then given the choice of rejecting the belief or the definition.

5. Usually, the respondent rejects the definition, because the belief is too central—epistemically or existentially—to be given up.

6. A new definition is offered, and the steps are repeated.

Research interviews today tend to be much less agonistic than this; the interview subject is commonly regarded as an informant or a partner, not as an opponent. The interviewer poses questions in order to obtain knowledge about the interviewee's world, and rarely enters into tenacious arguments about the logic and truth of what the interviewee says. Moreover, it is normally outside the scope of research interviews for the interviewer to argue the strength of his or her own conception of the topic investigated, or to try to change the subject's convictions. The interviewer is generally conceived as receptive rather than assertive (Wengraf, 2001).

From a philosophical perspective, we shall now ask more specifically how we may understand the kinds of knowledge that conversations can produce. Most of the knowledge produced in interview research can be said to be about people's experiences, desires, and opinions. To use a word from classical Greek philosophy, this kind of knowledge represents *doxa*. That is, it is about the interview subjects' experiences and opinions, which are often very interesting and important to learn about, but which—when viewed through the lenses of classical philosophy— rarely constitute knowledge in the sense of *episteme* (i.e., knowledge that has been found to be valid through conversational and dialectical questioning).

The purpose of the Socratic dialogues was to move the conversation partners from *doxa* to *episteme* (i.e., from a state of being simply *opinionated* to being capable of *questioning* and *justifying* what they believed to be the case) (Brinkmann, 2007a). Thus Socrates demonstrated that Agathon's opinion of the nature of love was unjustifiable—it was *doxa* rather than *episteme*—and Agathon had to admit that he did not know what he was talking about. If we follow Socrates, qualitative interviews seem to have the potential of being both doxastic and also epistemic; that is, they can elicit important descriptions and narratives of people's experiences, narratives, hopes, and dreams (the *doxa*), but they can also be employed as conversational ways of producing *episteme*, knowledge that has been justified discursively in a conversation.

The interviews conducted by Robert Bellah and his colleagues, as reported in *Habits of the Heart* (Bellah et al., 1985), which we return to a number of times in this book, come close to an epistemic interview form in the classical Socratic sense. In the appendix to their study of North American values and character, the authors spell out their view of social science and its methodology, summarized as "social science as public philosophy." The empirical material for their book consisted of interviews with more than 200 participants, some of whom were interviewed more than once. Inspired by Socratic dialogues, the researchers engaged in what they termed active interviews with their respondents in order to generate public conversation about societal values and goals. Such active interviews do not necessarily aim for agreement between interviewer and interviewee, and the interviewer is allowed to question and challenge what the interviewee says. In one example from their book, the interviewer tries to discover at what point the respondent would take responsibility for another human being:

Q: So what are you responsible for?

A: I'm responsible for my acts and for what I do.

Q: Does that mean you're responsible for others, too?

A: No.

Q: Are you your sister's keeper?

A: No.

Q: Your brother's keeper?

A: No.

Q: Are you responsible for your husband?

A: I'm not. He makes his own decisions. He is his own person. He acts his own acts. I can agree with them or I can disagree with them. If I ever find them nauseous enough, I have a responsibility to leave and not deal with it any more.

Q: What about children?

A: I . . . I would say I have a legal responsibility for them, but in a sense I think they in turn are responsible for their own acts. (Bellah et al., 1985, p. 304)

Here, the interviewer repeatedly challenges the respondent's claim of not being responsible for other human beings. With the Socratic principles of interviewing in mind, we can see the interviewer pressing for a contradiction between the respondent's definition of responsibility, involving the idea that she is only responsible for herself, and her likely feeling of at least some (legal) responsibility for her children. The individualist notion of responsibility is almost driven *ad absurdum,* but her restricted definition of responsibility apparently plays such a central role in the person's life that she is unwilling to give it up. It can be argued that this active and Socratic way of interviewing gives us important knowledge *primarily* about the doxastic individualist beliefs of Americans in the mid-eighties, and *secondarily* about the idea of responsibility in a normative-epistemic sense. For most readers would appreciate the above sequence as an argument that the respondent is wrong—she *is* responsible for other people, most clearly her children. At the very least, the reader is invited into an epistemic discussion, not just about private beliefs, but also about citizenship, virtue, responsibility, and ethics. The authors of *Habits of the Heart* conclude that unlike "poll data" generated by fixed questions that "sum up the *private* opinions," active (and in our terminology, epistemic) interviews "create the possibility of *public* conversation and argument" (Bellah et al., 1985, p. 305).

In the introductory chapter to this book, we presented an interview sequence from Bourdieu's (1999) *The Weight of the World.* Although the theme under discussion there was not a universal philosophical issue such as justice or virtue, we clearly see that Bourdieu as the questioner critically challenges the young men's account. As in some of the Socratic dialogues, and the interviews done in Bellah's study, this conversation approaches the form of a legal interrogation (Bourdieu confronts the respondents, as in these examples:

"You are not telling me the whole story . . ."; "But that took place during the daytime, at night?"). The study reported in *The Weight of the World* can be taken as an indication that epistemic interviews need not be limited to conceptual interviews or "elite interviews," like Socrates' conversations with the citizens of Athens, for "nonelites" are often capable of justifying their opinions and beliefs if challenged, and important knowledge sometimes develops from challenging respondents to give good reasons (see also the excerpt from the study of grading discussed in the first chapter).

We have not introduced the distinction between *doxa* and *episteme* in order to argue that only one of these should be sought in qualitative interviews. On the contrary, we believe that both can favorably be pursued in interview practices, according to one's knowledge interests, but it is probably fair to say that there has been in many interview studies a bias toward doxastic interviewing as the best way to conduct an interview. That is, texts on qualitative interviewing frequently regard the use of confronting questions that ask respondents to give reasons or reflect on abstract issues as an unfruitful way of conducting qualitative interviews. We now leave the philosophical discussion of *episteme* and epistemic interviews, but we return to the active, confronting interview style as an alternative to more empathetic and accepting interview styles in Chapter 8 on interview variations.

THERAPEUTIC INTERVIEWS AND RESEARCH INTERVIEWS

We shall now compare and contrast the research interview with the emotional personal interaction of therapeutic interviewing.

Carl Rogers was a therapist who pioneered the development of an open, client-centered interview form, originally termed "non-directive" (Rogers, 1945) and later changed to "client-centered," with the insight that all interviewing implies a sense of direction. Rogers was critical of Freud's theories of the unconscious and the speculative interpretations of psychoanalysis. Although the theoretical conceptions of client-centered therapy and psychoanalysis differ strongly, the client-centered interview in Box 2.5 comes fairly close to psychoanalytic interview practice. This concerns the intense emotional interaction as well as the few and cautious responses of the therapist. The interview was conducted by a counselor applying Rogers's therapeutic interview technique, an approach that was an inspiration for the early use of

---

BOX 2.5   A Therapeutic Interview on Hate

Subject: (*Silent for two minutes. Does not look at counsellor.*) You feel I want to come, but I *don't!* I'm not coming any more. It doesn't do any good. I don't like you. I hate you! I wish you never were born.

Counsellor: You just hate me very bitterly.

S: I think I'll throw you in the lake. I'll cut you up! You think people like you, but they don't . . . You think you can attract women, but you *can't* . . . I wish you were *dead*.

C: You *detest* me and you'd really like to get rid of me.

S: You think my father did bad things with me, but he didn't! You think he wasn't a good man, but he *was*. You think I want intercourse, but I *don't*.

C: You feel I *absolutely misrepresent* all your thoughts.

   . . .

S: You think I haven't been straight, but I have. I *hate* you. All I've had is pain, pain, pain. You think I can't direct my own life, but I can. You think I can't get well, but I can. You think I had hallucinations, but I didn't. I hate you. (*Long pause. Leans on desk in strained, exhausted pose.*) You think I'm crazy, but I'm not.

C: You're sure I think you're crazy.

S: (*Pause.*) I'm tied, and I just can't get loose! (*Despairing tone of voice, and tears. Pause.*) I had a hallucination and I've *got* to get it out!

   . . .

S: I knew at the office I had to get rid of this somewhere. I felt I could come down and tell you. I knew you'd understand. I couldn't say I hated myself. That's true but I couldn't say it. So I just thought of all the ugly things I could say to you instead.

C: The things you felt about yourself you couldn't say, but you could say them about me.

S: I know we're getting to rock bottom . . .

SOURCE: *Client-Centered Therapy* (pp. 211–213), by C. Rogers, 1956, Cambridge, MA: Houghton Mifflin.

qualitative research interviews (see Rogers, 1945, on the non-directive approach as a method for social research, allowing respondents to express themselves freely in the company of an accepting and empathetic researcher).

In this session, the client takes the lead right from the start, introduces the theme that is important to her—the detestable counselor—and expresses how much she hates him. He responds by reflecting and rephrasing her statements, emphasizing their emotional aspects. He does not, as would be likely in a normal conversation, take issue with the many accusations against him. In this sequence the counselor does not ask questions for clarification, nor does he offer interpretations. At the end, after "she has got it all out," the client acknowledges the counselor's ability to understand her, and she herself offers an interpretation: I couldn't say I hated myself, so I just thought of all the ugly things I could say to you instead. We may note that the counselor's interventions were not entirely non-directive; the client introduces several themes— such as not wanting to come to therapy, it does not do her any good and objecting to the therapist's belief that her father did wrong things with her— whereas the counselor consistently repeats and condenses her negative statements about himself, which lead the client to an emotional insight about her self-hatred.

A therapeutic interview aims at change through an emotional personal interaction rather than through the logical argumentation used in a philosophical dialogue. The changes sought are not primarily conceptual, but emotional and personal. Although the main purpose of therapeutic interviews is to assist patients in overcoming their suffering, a side effect has been the production of knowledge about the human situation. Both a therapeutic and a research interview may lead to increased understanding and change, but the emphasis is on knowledge production in a research interview and on personal change in a therapeutic interview.

Although Carl Rogers and Sigmund Freud had different theories of human personality and therapy, with Rogers emphasizing the present and conscious experience and Freud the past and the unconscious, their therapeutic practices were in several ways rather close. Thus the emotional therapeutic session above could also have been part of a psychoanalytic session. The psychoanalytic interview, where knowledge production is not the primary purpose, has been *the* psychological method for providing significant new knowledge about humankind. Freud regarded the therapeutic interview as a research method: "It is indeed one of the distinctions of psychoanalysis that research and treatment proceed hand in hand" (1963, p. 120).

BOX 2.6   The Psychoanalytic Research Interview

The Individual Case Study. Psychoanalytic therapy is an intensive case study of individual patients over several years. The extensive knowledge of the patient's life world and of his or her past thereby obtained provides the therapist with a uniquely rich context for interpreting the patient's dreams and symptoms.

The Open Mode of Interviewing. The psychoanalytic interview takes place in the structured setting of the therapeutic hour, the content is free and non-directive; it is based on psychoanalytic theory, yet proceeds in an open manner. To the patient's free associations corresponds the therapist's "evenly-hovering attention." Freud warned against formulating a case scientifically during treatment, since it would interfere with the open therapeutic attitude in which one proceeds "aimlessly, and allows oneself to be overtaken by any surprises, always presenting to them an open mind, free from any expectations" (Freud, 1963, p. 120).

The Interpretation of Meaning. An essential aspect of psychoanalytic technique is the interpretation of the meaning of the patient's statements and actions. The psychoanalytic interpretations are open to ambiguity and contradictions, to the multiple layers of meaning of a dream or a symptom. They require an extensive context, with the possibility of continual reinterpretations: "The full interpretation of such a dream will coincide with the completion of the whole analysis: if a note is made of it at the beginning, it may be possible to understand it at the end, after many months" (Freud, 1963, p. 100).

The Temporal Dimension. Psychoanalytic therapy unfolds over several years and thus has a historical dimension, with a unique intertwinedness of the past, present, and future. Freud's innovation was here to see human phenomena in a meaningful historical perspective; the remembrance of the past is an active force of therapeutic change, and the therapy aims at overcoming the repressions of the past and the present resistance toward making the unconscious conscious.

The Human Interaction. Psychoanalytic therapy takes place through an emotional human interaction, with a reciprocal personal involvement. Freud noticed that if the analyst allows the patient time, devotes serious interest to the patient, and acts with tact, a deep attachment of the patient to the therapist develops. The strong emotions, ranging from love to rage, are interpreted as "transference" of childhood feelings for the parents to the therapist. This transference is deliberately employed by the therapist as a means to overcome the patient's emotional resistance toward a deeper self-knowledge and change. Different depths of layers of the patient's personality are disclosed, depending on the intensity of the patient's emotional ties to the therapist. The transference of the therapist's own feelings to the patient, termed "counter-transference," is not eliminated but employed in the therapeutic process as a reflected subjectivity.

Pathology as Topic of Investigation. The subject matter of psychoanalytic therapy is the abnormal and irrational behavior of patients in crisis, their apparently meaningless and bizarre symptoms and dreams. The pathological behavior serves as a magnifying glass for the less visible conflicts of average individuals; the neuroses and psychoses are extreme versions of normal behavior, which are the characteristic expressions of what has gone wrong in a given culture.

The Instigation of Change. The mutual interest of patient and therapist is to overcome the patient's suffering from neurotic symptoms. Despite patients having sought treatment voluntarily, they exhibit a deeply seated resistance to a change in self-understanding and action. "The whole theory of psychoanalysis is . . . in fact built up on the perception of the resistance offered to us by the patient when we attempt to make {the patient's] unconscious conscious" (Freud, 1963, p. 68). While understanding may lead to change, the implicit theory of knowledge in psychoanalysis is that a fundamental understanding of a phenomenon can be obtained by attempting to change the phenomenon.

Box 2.6 shows seven characteristics of the psychoanalytic interview based on Freud's writings on the therapeutic technique (see Kvale, 2003). While main features of the psychoanalytic interview are ethically off limits for research interviewing, contemporary interview researchers may still learn from this and other therapeutic forms of interviewing. The psychoanalytic interview is related to, but also contrasts with, the research interview and its mode of understanding. The purpose of a therapeutic interview is the facilitation of changes in the patient, and the knowledge acquired from the individual patient is a means for instigating personality changes. The general knowledge of the human situation gained through the psychoanalytic process is a side effect of helping patients overcome their neurotic suffering. The intensive personal therapeutic relationship may open painful, hidden memories and deeper levels of personality, which are inaccessible through a brief research interview. In a qualitative research interview, the purpose is to obtain knowledge of the phenomena investigated and any change in the interviewed subject is a side effect.

There are many problems with psychoanalysis as a research method, and the scientific status of psychoanalytic knowledge is contested (see, e.g., Fisher & Greenberg, 1977). Yet, it is a continuing paradox that the therapeutic interview, which has not been accepted as a scientific method and for which general knowledge production is a side effect, has produced some of the most viable knowledge in the discipline of psychology. Psychoanalysis is the one branch of psychology that, more than a century after its inception, still has a strong professional impact on psychotherapy and continues to be of interest to the general public, to other sciences, and to philosophers. Central areas of current psychology textbooks are based on knowledge originally obtained through the psychoanalytic interview regarding dreams and neuroses, sexuality, childhood development and personality, anxiety and motivation, and the unconscious forces.

Despite the significant knowledge production of psychoanalytic therapy, in textbooks of psychological methods, the major method by which psychoanalytic knowledge is obtained—the psychoanalytic interview—is absent. Though generally critical of the speculative and reductionist trends of psychoanalytic theory, philosophers have reflected on the unique nature of the personal interaction in the psychoanalytic interview and its potentials for personal change as well as its contributions to knowledge about the human situation. Among the philosophical texts addressing psychoanalysis are Sartre's (1963) existential mediation on psychoanalysis and Marxism in *The Problem of Method;* Ricoeur's (1970) phenomenological and hermeneutical *Freud and Philosophy: An Essay on Interpretation;* and Habermas's (1971) critical hermeneutical analysis of psychoanalysis as a model for an emancipatory social science in *Knowledge and Human Interests.*

Despite the radical differences between research interviews and psychoanalytic interviews—ethically and methodologically, it is possible for research interviewers to learn from the modes of questioning and interpreting developed in therapeutic interviews. The development of the free association interview by Hollway and Jefferson (2000) is a recent case in point. These researchers argue that qualitative interviewers always have an explicit or implicit theory of the subject, and their theory is based on the psychoanalytic idea of "the defended subject." They believe that "subjects are motivated *not* to know certain aspects of themselves and . . . they *produce* biographical accounts which avoid such knowledge" (p. 169). Thus, in order to interpret the subjects' free associations, researchers should be familiar with psychoanalytic theory.

In the preceding chapter we mentioned the influence of the psychoanalytic interview on the interviewing techniques of Piaget and the Hawthorne studies, as well as on the motivational market interviews, which also found inspiration in Rogers's non-directional interviewing. It was the psychologist Elton Mayo who developed the sophisticated method of interviewing used in the Hawthorne studies.

BOX 2.7   Elton Mayo's Method of Interviewing

1. Give your whole attention to the person interviewed, and make it evident that you are doing so.

2. Listen—don't talk.

3. Never argue; never give advice.

4. Listen to:

   (a) what he wants to say

   (b) what he does not want to say

   (c) what he cannot say without help

5. As you listen, plot out tentatively and for subsequent correction the pattern (personal) that is being set before you. To test this, from time to time summarize what has been said and present for comment (e.g., "is this what you are telling me?"). Always do this with the greatest caution, that is, clarify in ways that do not add or distort.

6. Remember that everything said must be considered a personal confidence and not divulged to anyone.

SOURCE: *The Social Problems of an Industrial Civilization* (p. 65), by E. Mayo, 1933, New York: MacMillan.

Mayo's approach to interviewing, outlined in Box 2.7, was much inspired by psychoanalytic therapeutics and an emerging emotional ethos (see Illouz, 2007), and his recommendations for interviewers prove to be surprisingly contemporary. Mayo's method of interviewing could, without much change, appear in most introductory books on qualitative interviewing today.

In this chapter, we have attempted to exemplify and outline the mode of understanding of a phenomenological life world interview. We have further drawn in the philosophical dialogue and the therapeutic interview as related but contrasting forms of interviewing, and pointed to their relevance for understanding current research interviewing. We shall now turn to the nature of the knowledge produced by research interviews.

THREE

EPISTEMOLOGICAL
ISSUES OF INTERVIEWING

—•◦•—

Q ualitative research can give us compelling descriptions of the qualitative human world, and qualitative interviewing can provide us with well-founded knowledge about our conversational reality. Research interviewing is thus a knowledge-producing activity, but the question is how to characterize the form of knowledge that qualitative research interviewing can give us. In this chapter we address epistemological issues of research interviewing. *Epistemology* is the philosophy of knowledge and involves long-standing debates about what knowledge is and how it is obtained. Throughout this book we show how epistemological presumptions of qualitative interview knowledge concretely bear upon conceiving and practicing research interviewing. This concerns, for example, issues such as whether an interview subject's spontaneous narratives are to be regarded as digressions from the scientific task of finding facts and whether narratives are essential aspects of human knowledge.

We first propose two metaphors, the interviewer as a miner and the inter-viewer as a traveler, and go on to discuss the knowledge produced in interviews in relation to conceptions of knowledge in a postmodern age, drawing upon postmodern thought as well as hermeneutic and pragmatic conceptions of knowledge. Inspired by these epistemological positions, we depict seven key features of interview knowledge as produced, relational, conversational, con-textual, linguistic, narrative, and pragmatic. We end the chapter by outlining positivist philosophy, which portrays research as rule governed and scientific knowledge as quantitative and has served to rule out qualitative interviewing as a legitimate research method.

THE INTERVIEWER AS A MINER OR AS A TRAVELER

These two contrasting metaphors of the interviewer—as a miner or as a traveler— illustrate the different epistemological conceptions of interviewing as a process of *knowledge collection* or as a process of *knowledge construction,* respectively. By *metaphor,* we refer to understanding one kind of thing by means of another, thereby highlighting possible new aspects of a kind. The two metaphors for interviewing, although not logically distinct categories, may inspire the researcher to reflect upon what conceptions of knowledge he or she brings to an interview inquiry.

In a *miner metaphor,* knowledge is understood as buried metal and the interviewer is a miner who unearths the valuable metal. The knowledge is waiting in the subject's interior to be uncovered, uncontaminated by the miner. The interviewer digs nuggets of knowledge out of a subject's pure experiences, unpolluted by any leading questions. The nuggets may be understood as objective real data or as subjective authentic meanings. A research interviewer strips the surface of conscious experience, whereas a therapeutic interviewer mines the deeper unconscious layers. The knowledge nuggets remain constant through transcription from an oral conversation to a written transcript. By means of a variety of data-mining procedures, the researcher extracts the objective facts or the essential meanings, today preferably by computer programs.

As an alternative, in the *traveler metaphor* the interviewer is a traveler on a journey to a distant country that leads to a tale to be told upon returning home. The interviewer-traveler wanders through the landscape and enters into conversations with the people he or she encounters. The traveler explores the many domains of the country, as unknown terrain or with maps, roaming freely around the territory. The interviewer-traveler, in line with the original Latin meaning of *conversation* as "wandering together with," walks along with the local inhabitants, asking questions and encouraging them to tell their own stories of their lived world; some, such as the anthropologists, living for a longer time with their conversation partners. The potentialities of meanings in the original stories are differentiated and unfolded through the traveler's interpretations of the narratives he or she brings back to home audiences. The journey may not only lead to new knowledge; the traveler might change as well. The journey might instigate a process of reflection that leads the traveler to new ways of self-understanding, as well as uncovering previously taken-for-granted values and customs in the traveler's home country.

These two metaphors for the interviewer—as a miner and as a traveler—represent contrasting ideal types of interview knowledge as respectively given or constructed. The two metaphors stand for alternative genres and have different rules of the game. A miner approach will tend to regard interviews as a site of data collection separated from the later data analysis. A traveler conception leads to interviewing and analysis as intertwined phases of knowledge construction, with an emphasis on the narrative to be told to an audience. The data-mining conception of interviewing is close to the mainstream conception of modern social sciences where knowledge is already there, waiting to be found, whereas the traveler conception is nearer to anthropology and a postmodern constructive understanding that involves a conversational approach to social research.

We should note that the miner metaphor pertains not only to positivist and empiricist data collection, but also to a certain extent to Socrates' pursuit for preexisting truths, to Husserl's search for phenomenological essences, and to Freud's quest for hidden meanings buried in the unconscious (see his archae-ology metaphor for the psychoanalytic excavations of the unconscious). Some traditions may imply both metaphors, such as psychoanalysis, where the nar-rative constructions of case histories come closer to a traveler conception of knowledge. We may further, inspired by Bauman (1996), discern two types of travelers: the pilgrim on a long search for truth and the tourist shopping for experiences. The pilgrim's goal is set according to shared external standards (e.g., concerning how to live an ethical life based on God's command), whereas the tourist invents his or her own goals according to aesthetic criteria based on taste and lifestyle. In a postmodern consumer society, aesthetic cri-teria concerning beauty, ethical criteria concerning goodness, and political cri-teria concerning justice all compete with epistemic criteria concerning truth. In sum, the miner and the traveler metaphors may, in a simple dichotomized form, illustrate the complex and contested conceptions of interview knowl-edge. Below, we turn to more sophisticated epistemological conceptions of knowledge as they pertain to research interviews.

When discussing the epistemology of interviewing, it should be kept in mind that the interview is a special form of conversational practice, which was developed in everyday life over centuries in relative independence from epistemological discussions. In the last few centuries, interviews have become institutionalized as various forms of professional interviews. Although the varying forms of interviews have not been developed from any specific theory or epistemological paradigm, we may, however, post hoc, invoke different epistemological positions to conceptualize the knowledge that is produced in interviews. A clarification of such positions may serve to shed light on different understandings and practices of research interviewing.

INTERVIEWS IN A POSTMODERN AGE

Different philosophies highlight different aspects of knowledge relevant to the qualitative interview. In this chapter the emphasis is on knowledge and interviews in a postmodern age, with a focus on hermeneutics, pragmatism, and, in particular, postmodern thought. While some of their epistemological assumptions of knowledge differ, as do their geographical birthplaces (postmodernism is associated with French thinkers, pragmatism with Americans, and hermeneutics with German philosophers), they may here serve as contexts for reflection on the multiple aspects of producing knowledge through interviews. Qualitative research interviewing has existed in the social sciences for nearly a century, but it did not become a general issue for methodological discussions until the last few decades. This may in part be due to social scientists not having had access to philosophies relevant to conceptualize the kind of knowledge produced by research interviewing. In the following section we argue that the philosophical positions mentioned above may provide conceptual frames of reference, which may clarify the nature and the problems, the strengths and the weaknesses, of knowledge produced by qualitative research interviews.

Hermeneutics is the study of the interpretation of texts. From a hermeneutical viewpoint, the interpretation of meaning is the central theme, with a specification of the kinds of meanings sought and attention to the questions posed to a text. The concepts of conversation and of text are pivotal in the hermeneutical tradition in the last centuries of the humanities, and there is an emphasis on the interpreter's foreknowledge of a text's subject matter. The purpose of hermeneutical interpretation is to obtain a valid and common understanding of the meaning of a text. Although the subject matter of classical hermeneutics was the texts of religion, law, and literature, there has been an extension of the concept of "text" to include discourse and even action. Thus, in *Truth and Method,* Gadamer (1975) begins with Plato's dialogues and regards both the conversation and the oral tradition as presuppositions for understanding the written texts, which historically are secondary phenomena. According to Gadamer, we are conversational beings for whom language is a reality (see Bernstein, 1983). In his article "The Model of the Text: Meaningful Action Considered as a Text," Ricoeur (1971) extends the hermeneutic principles of interpretations of the texts of the humanities to the interpretation of the object of the social sciences—meaningful action. Human beings are self-interpreting, historical creatures, whose means of understanding are provided by tradition and historical life. Understanding depends on certain *pre-judices,* as Gadamer famously argued. And every text derives its meaning from a con-text. Knowledge of what others are doing and saying, of what their actions and utterances mean, always depends "upon some background or context of other meanings, beliefs, values, practices, and so forth." (Schwandt, 2000, p. 201). From hermeneutics, qualitative researchers can learn to analyze their interviews as texts and look beyond the here and now in the interview situation, for example, and pay attention to the contextual interpretive horizon provided by history and tradition (see Palmer, 1969).

Pragmatism as a philosophical position, with its central view that language and knowledge do not copy reality but are means of coping with a changing world, has come to the fore in a postmodern age. Pragmatism emphasizes the primacy of practice and the use-value of the ideas and theories produced by researchers. Pragmatism was originally developed by American philosophers such as Peirce, James, and Dewey in the transition from the 19th to the 20th century, and is today represented by Rorty and Putnam, among many others. In Rorty's neopragmatic philosophy, conversation is a basic mode of knowing: "We see knowledge as a matter of conversation and of social practice, rather than as an attempt to mirror nature" (Rorty, 1979, p. 171). From pragmatism, interview researchers can learn to focus on the practical aspects of what they are doing, on the craftsmanship of their activities, and on the issues of values and ethics raised by the use-value of their research results.

In the pragmatic approach of the present book, the emphasis is less on paradigmatic legitimation of interview research than on the practical implications of the different epistemological positions for the craft of research interviewing. In later chapters, we give examples of how different epistemological positions lead to different conceptions of interview research, and also to different forms of practice, not least concerning the many decisions about how to do it that are made throughout an interview investigation. This concerns issues such as the use of leading questions, the nature of transcriptions, forms of interview analysis, and also the understanding of objectivity and validity of interview knowledge.

In addition to hermeneutics and pragmatism, a *phenomenological* perspective and a *dialectical* approach are important philosophical positions in relation to qualitative interviewing. *Phenomenology* was treated in the previous chapter on interview conversations and includes a focus on consciousness and the life world, an openness to the experiences of the subjects, a primacy of precise descriptions, attempts to bracket foreknowledge, and a search for invariant essential meanings in the descriptions. A *dialectical* standpoint focuses on the contradictions of a statement and their relations to the contradictions of the social and material world. There is an emphasis on the new, rather than on the status quo, and on the intrinsic relation of knowledge and action. A dialectical position will be brought up in relation to discursive analyses of interviews in Chapter 13.

In *postmodern thought,* there is a disbelief in universal systems of thought (Lyotard, 1984). There is a lack of credibility of meta-narratives of legitimation—such as the Enlightenment belief in progress through knowledge and science. The modern conception of knowledge as a mirror of reality is replaced by a conception of the social construction of reality, where the focus is on the interpretation and negotiation of the meanings of the social world. With the breakdown of the universal meta-narratives of legitimation, there is an emphasis on the local contexts, on the social and linguistic construction of a social reality where knowledge is validated through practice. There is openness to qualitative diversity, to the multiplicity of local meanings; knowledge is perspectival, dependent on the viewpoint and values of the investigator. With a decline of modern universal systems of knowledge, the narratives of local, manifold, and changing language contexts come into prominence. The linguistic turn in philosophy has been radicalized in postmodern philosophy: In some versions of postmodernism, language constitutes reality, each language constructing reality in its own way. The focus on language shifts attention away from the notion of an objective reality, and also away from the individual subject. There is no longer a unique and sovereign self who uses language to describe an objective world or to express itself; it is the structures of language that speak through the person. In *The Postmodern Condition,* Lyotard (1984) also depicted economic performativity, striving for the most efficient input-output ratio, as crucial for knowledge in a postmodern age. Knowledge is increasingly understood as a commodity, a tendency Lyotard calls the mercantilization of knowledge.

In a postmodern epistemology, the certainty of our knowledge is less a matter of interaction with a nonhuman reality than a matter of conversation between persons. Knowing subjects are conceived not as isolated islands but as existing in "a fabric of relations" (Lyotard, 1984, p. 15). Knowledge is neither inside a person nor outside in the world, but exists in the relationship between persons and world. Merleau-Ponty, a phenomenological psychologist and philosopher whose work has also been regarded as a precursor to post-modern thought, concludes his *Phenomenology of Perception* (1962) with a quote from Saint Exupery: "Man is but a network of relations." In postmodern epistemology, there is a shift from the individual mind to relations between persons: "Constructionism replaces the individual with the relationship as the locus of knowledge" (Gergen, 1994, p. x).

Leading postmodern theorists in the second half of the 20th century were French philosophers such as Lyotard, Baudrillard, Derrida, and Foucault, although not all of these identify themselves explicitly with postmodernism. A postmodern approach to interviewing focuses on the interview as a production site of knowledge, on its linguistic and interactional aspects, including the differences between oral discourse and written text, and emphasizes the narratives constructed in the interview. See Rosenau (1992) and Scheurich (1997) for broader discussions of postmodern approaches to the social sciences, and, for a pertinent overview of philosophical positions and issues relevant to qualitative research, see Schwandt (2001).

SEVEN FEATURES OF INTERVIEW KNOWLEDGE

With inspiration from the philosophical conceptions depicted above, we shall now describe interview knowledge with respect to seven key features. Interview knowledge is produced, relational, conversational, contextual, linguistic, narrative, and pragmatic. These intertwined features are taken as a starting point for clarifying the nature of the knowledge yielded by the research interview and for developing its knowledge potential. These features are characteristic not only of interview knowledge, but also of the objects that interviews are able to give us knowledge about. That is, the lived social and historical world of human interaction is itself something constantly produced by humans; it is also relational, conversational, contextual, linguistic, narrative, and pragmatic or action oriented. Throughout the discussions of the seven practical stages of an interview project in Part II of this book, we draw in the aspects of knowledge outlined below.

Knowledge as Produced. The research interview is a production site of knowledge. Interview knowledge is socially constructed in the interaction of interviewer and interviewee. The knowledge is not merely found, mined, or given, but is actively created through questions and answers, and the product is co-authored by interviewer and interviewee. The production process continues through the transcription, analysis, and reporting of the original interviews, with the reported knowledge tinged by the procedures and techniques applied on the way.

Knowledge as Relational. The knowledge created by the inter-view is interrelational and inter-subjective. As illustrated by the ambiguous vase/faces in Figure 1.1, the researcher can focus on the knowledge produced inter the views of the interviewer and interviewee or concentrate on the interaction between the two participants. Therapists have been attentive to the interpersonal relationships in their interviews. A therapeutic interview is thus an inter-personal situation where the data produced are neither objective nor subjective, but intersubjective (Sullivan, 1954). The research interview establishes new relations in the human webs of interlocution, with the goal of producing knowledge about the human situation.

Knowledge as Conversational. With the loss of faith in an objective reality that can be mirrored and mapped in scientific models, attention must be paid to discourse and negotiation about the meaning of the lived world. Philosophical discourse and research interviews rely on conversations giving access to knowledge. Also, in the classical philosophical position of Socrates, conversations are a primary way of producing knowledge about the true, the good, and the beautiful. If we follow Socrates, we understand qualitative interviews as having the potential of producing descriptions and narratives of everyday experiences as well as the epistemic knowledge justified discursively in a conversation.

Knowledge as Contextual. Hermeneutic philosophy has emphasized the fact that human life and understanding is contextual, both in the here and now and in a temporal dimension. Knowledge obtained within one situation is not automatically transferable to, nor commensurable with, knowledge within other situations. The interview takes place in an interpersonal context, and the meanings of interview statements relate to their context. Interviews are sensitive to the qualitative differences and nuances of meaning, which may not be quantifiable and commensurable across contexts and modalities. When it comes to ethical judgments of an interview procedure, and qualitative analytical generalizations of the knowledge produced, thick contextual descriptions of the settings are required. With the heterogeneity of contexts, the issue of translation between contexts comes into the foreground, for example from the interviewers' conversations with their subjects to their conversations with other researchers about the validity of the interview knowledge produced, and also when the results of the conversations enter into a public conversation about the knowledge produced.

Knowledge as Linguistic. Language is the medium of interview research; language is the tool of the interview process, and the resulting interview product is linguistic in the form of oral statements and transcribed texts to be analyzed. The transition from one linguistic modality to another, such as from oral to written language, is not merely a technical question of transcription, but raises issues concerning the different natures of oral and written language. Knowledge is constituted through linguistic interaction, and the participants' discourses and their effects are of interest in their own right. A variety of approaches exist for analyses of interviews that are based on language, such as linguistic, conversational, narrative, discursive, and deconstructive analyses.

Knowledge as Narrative. Stories are a powerful means of making sense of our social reality and our own lives. The interview is a key site for eliciting narratives that inform us of the human world of meanings. In open interviews, people tell stories about their lives; see, for example, the phenomenological interview in Chapter 2, where the respondent spontaneously produced a narrative on her learning of interior design. Research interviews give access to the manifold local narratives embodied in storytelling and they may themselves be reported in a narrative form.

Knowledge as Pragmatic. When human reality is understood as conversation and action, knowledge becomes the ability to perform effective actions. Today, the legitimacy question of whether a study is scientific, or whether it leads to true knowledge, tends to be replaced by the pragmatic question of whether it provides useful knowledge. Good research is research that works. The issue concerning what should count as "useful" is laden with value and ethical questions, to which we turn in the following chapter on ethics. There is an insistence in pragmatism that ideas and meanings derive their legitimacy from enabling us to cope with the world in which we find ourselves.

KNOWLEDGE AND
INTERVIEWS IN A POSITIVIST CONCEPTION

We conclude this chapter with an outline of a philosophical tradition, which has been influential in the social sciences, and which has contributed to outlawing or marginalizing qualitative research as a legitimate scientific approach. A positivist philosophy has often been implied by researchers who are skeptical toward qualitative research interviewing, since positivism emphasizes the point that data should be quantitative. Scientific methods should further be neutral with regard to the subjectivity, interests, and values of the researcher. Because of the importance of positivism as an often-invoked antithesis to qualitative research, we briefly discuss it here by distinguishing between the classical positivism of Auguste Comte, which does not contradict the practice of qualitative interviewing, and the later restrictive methodological positivism of the social sciences.

A Rehabilitation of Classical Positivism?

Positivist philosophy undoubtedly made a historical contribution to the social sciences and also to the arts: Auguste Comte (1798–1857) founded both positivist philosophy and the science of sociology. The positivist philosophy reacted against religious dogma and metaphysical speculation and advocated a return to observable data. Positivist science was to provide determinate laws of society with the possibility of socially engineering society.

The influence of positivist sociology can be seen in the work of Émile Durkheim, an early sociologist who gave penetrating qualitative analyses of social phenomena. Positivism also had an extended influence on the arts of the 19th century, inspiring a move from mythological and aristocratic themes to a new realism, depicting in detail the lives of workers and the bourgeoisie (for some of this history, particularly in the British context, see Dale, 1989).

In histories of music, Bizet's opera *Carmen,* featuring the lives of cigarette smugglers and toreadors, is depicted as inspired by positivism, and Flaubert's realistic descriptions of the life of his heroine in *Madame Bovary* enable it to be considered a positivist novel. Impressionist paintings, sticking to the immediate sense impressions, in particular the sense data of pointillism, also drew inspiration from positivism. Michel Houllebecq is a contemporary French author who explicitly acknowledges his inspiration from Comte's positivism, and Houellebecq has written the preface for a recent volume on Comte today (Bourdeau, Braunstein, & Petit, 2003).

The early positivism was also a political inspiration for feminism, and it was the feminist Harriet Martineau who translated Comte's *Positive Philosophy* into English. In philosophy, the founder of phenomenological philosophy, Husserl, stated that if positivism means being faithful to the phenomena, then we, the phenomenologists, are the true positivists. It can even be argued that the insistence in Comte's positivism to stay close to observed phenomena rather than engaging in metaphysical speculation about theoretical entities comes close to a postmodern emphasis on the importance of staying close to observable surface phenomena rather than postulated deep structures—here the surface has become the essence.

Methodological Positivism

The open approach of classical positivism was lost in the methodological positivism of the Vienna circle in the 1920s, whose members included the philosophers Schlick, Carnap, and Neurath (see Radnitzky, 1970; Schwandt, 2001). Its strict focus on the logic and validity of scientific statements contributed to a methodological bureaucracy of social science research, particularly in the mid-century United States. Bureaucracy is characterized by standardized procedures and methods, regularity, formal rules of decision and impersonal impartiality, written communication, and quantification.

A rigorous positivist epistemology came to dominate social science textbooks on methodology from the middle of the 20th century. A "unity of science" was advocated, where scientific research was based on a common method, independent of the subject matter investigated. In methodological positivism, scientific knowledge was to be found by following general methodological rules that were largely independent of the content and context of the investigation. The nature of scientific methods was to be found in the advanced natural sciences. Thus scientific statements should be based upon observable data; the observation of the data and the interpretation of their meanings were to be strictly separated. Scientific facts were to be unambiguous, intra-subjectively and inter-subjectively reproducible, objective, and quantifiable. Scientific statements ought to be value neutral, facts were to be distinguished from values, and science from ethics and politics. Any influence of the subjectivity of the researcher should be eliminated or minimized.

According to an epistemology that takes as its starting point the elimination of human subjectivity in research, the qualitative interview based on interpersonal interaction is unscientific. Interview data consist of meaningful statements, themselves based on interpretations; the data and their interpretations are thus not strictly separated. Quantified knowledge is not the goal of interview research; interview findings are commonly expressed in language, frequently in everyday language. Interview statements can be ambiguous and contradictory and the findings may not be intersubjectively reproducible, for example, because of the interviewers' varying knowledge of and sensitivity to the interview topic. In conclusion, major features of the mode of understanding in the qualitative interview appear, from a methodological positivist perspective, as sources of error, and the interview, following a positivist perspective, therefore cannot be a scientific method.

Although social scientists have often labeled positivist research as uncritical, since it regards critiques of the historical and social functions of social research as outside the scientific domain, it should be kept in mind that the positivists in fact contributed to moving social research beyond myth and common sense. Their emphasis on using and reporting transparent methods for arriving at scientific data opens the possibility of intersubjective control and critiques of research findings, counteracting subjective and ideological bias in research.

Critiques of positivism in social science are today often dismissed as attacking a man of straw. A strict methodological positivist epistemology is rarely if ever advocated by philosophers of science today. Within the social sciences, however, the formal methodological rules of positivist science still prevail in certain places—in newer neopositivist positions, in many mainstream methodology textbooks, and particularly in the new discourse on evidence-based practice, where evidence is frequently understood as based on formalized quantitative research.

BOX 3.1   Evidence-Based Practice

The approach of evidence-based practice was developed by the British epidemiologist Cochrane in the 1970s. Faced with the wealth of new bio-medical research and contradictions in this research, Cochrane sought to work out a model to evaluate rigor in medical research, which could support advice to practioners about which drugs and medical interventions had well-documented effectiveness. This endeavor led to an evidence hierarchy that placed randomized controlled experiments as "the gold standard," and expert opinion, as well as qualitative research, at the bottom level of evidence.

Such strict criteria of evidence are perhaps adequate for the biomedical research they were developed for. However, when they are extrapolated to other forms of research, they may result in a "politics of evidence" (Morse, 2006), where qualitative research in general becomes marginalized. The explorative, interactive, and case-based approach of many qualitative studies does not fit the logic of strictly controlled experimentation. In some areas, there have been attempts to broaden the original rigorous criteria—for example, evidence-based practice in psychology (EBPP), which aims for "the integration of the best available research with clinical expertise in the context of patient characteristics, culture, and preferences" (see the Report of the Presidential Task Force on Evidence-Based Practice, Levant, 2005). And while there are attempts to develop evidence criteria for qualitative research, the effect of the evidence-based practice movement on qualitative research has largely been to discredit qualitative research, hampering the acceptance of research proposals and the funding of qualitative research, and to support a methodological and political conservatism (Denzin & Giardina, 2006).

Box 3.1 depicts the role of the evidence-based practice movement in medical research and its relation to the legitimacy of qualitative research. We see here a parallel to methodological positivism, where generalized criteria developed for one area of scientific research—experimentation in physics—were extended to research in general, thereby outlawing qualitative social research. Today, a corresponding methodological imperialism takes place, where methodical criteria developed for evidence in biomedical research are generalized to the social sciences, again relegating qualitative research to an inferior position.

******

This chapter has treated the epistemology of qualitative interview research, primarily with reference to postmodern, pragmatist, and hermeneutic philosophies. We have argued that these philosophical positions can help us understand the nature of the knowledge produced by qualitative interviews, constituted through language, narrative, human relations, and contexts. We have also presented a positivist philosophy, which emphasizes research as rule governed and scientific knowledge as quantitative, that has served to rule out qualitative interviewing as a legitimate research method. We have emphasized that in contrast to methodological positivism, knowledge is not obtained in qualitative research by following value- and interest-free methods, for the subjectivities of human beings play an irreducible role in qualitative knowledge production. Rather than excellence in research being conceptualized in terms of the methods used, we will advocate in Chapter 5 that excellent qualitative research is marked by good craftsmanship. This theme is also pursued in the next chapter, on ethics, where we address some of the ethical uncertainties of interviewing as a social practice. A central point will be that practicing ethically capable research cannot be reduced to following ethical principles and guidelines but must include elements of situated human judgment.

⊰ FOUR ⊱

ETHICAL ISSUES
OF INTERVIEWING

————•◦◆◦•————

I n this chapter we discuss some of the ethical or moral concerns that are
involved in the practice of interviewing for research purposes. We aim to
show how ethical issues go beyond the live interview situation and are embedded in all stages of an interview inquiry. We address ethical guidelines for
social research and the importance of informed consent, confidentiality, consequences, and the researcher's role. In the concluding chapter, Chapter 17, we
return to the moral issues of interviewing and go beyond the microethics of an
interview project to address the macroethics of the broader social effects of the
interview-produced knowledge.

In line with the practical point of view of this book, where we emphasize
the craftsmanship of qualitative research, we view moral conduct as closely
connected to the practical skills of situated judgment, which Aristotle (1994)
described as *phronesis,* the intellectual virtue of recognizing and responding to
what is most important in a situation. As a consequence, the practical skills of
the interview researcher, which enable him or her to understand the concrete
powers and vulnerabilities that are in play in particular situations, come into
focus. We return to this issue in the final sections of this chapter, where we
address the issue of how interview researchers can be educated in order to deal
with concrete ethical problems.

INTERVIEWING AS A MORAL INQUIRY

An interview inquiry is a moral enterprise. Moral issues concern the means as well as the ends of an interview inquiry. The human interaction in the interview affects the interviewees, and the knowledge produced by an interview inquiry affects our understanding of the human condition. Consequently, interview research is saturated with moral and ethical issues. *Ethics* comes from the Greek *ethos,* which means character, and was translated into the Latin *mores* (from which we have *morality*), which also means character, custom, or habit (Annas, 2001). In this chapter, we do not distinguish systematically between *ethics* and *morality,* although the former term typically indicates something formal (e.g., "ethical guidelines"), whereas the latter term normally points to everyday conduct (e.g., "the morality of everyday life"). We work with a broad definition of these concepts to refer to *the oughtness of human existence,* that is, to the idea that human life involves moral demands to act, think, feel, and be in required ways.

The undertaking of a research project raises questions as to the value of the knowledge produced, concerning the social contributions of the study. Social science research should serve scientific *and* human interests. The preamble to an earlier version of the American Psychological Association's ethical principles thus emphasized that psychologists are committed to increasing knowledge of human behavior and of people's understanding of themselves and others, and to utilizing this knowledge for the promotion of human welfare: "The decision to undertake research rests upon a considered judgment by the individual psychologist about how best to contribute to psychological science and human welfare" (APA, 1981, p. 637).

ETHICAL ISSUES
THROUGHOUT AN INTERVIEW INQUIRY

Ethical problems in interview research arise particularly because of the complexities of "researching private lives and placing accounts in the public arena" (Birch et al. 2002, p. 1). Ethical issues go through the entire process of an interview investigation, and potential ethical concerns should be taken into consideration from the very start of an investigation to the final report.

BOX 4.1   Ethical Issues at Seven Research Stages

Thematizing. The purpose of an interview study should, beyond the scientific value of the knowledge sought, also be considered with regard to improvement of the human situation investigated.

Designing. Ethical issues of design involve obtaining the subjects' informed consent to participate in the study, securing confidentiality, and considering the possible consequences of the study for the subjects.

Interview Situation. The personal consequences of the interview interaction for the subjects need to be taken into account, such as stress during the interview and changes in self-understanding.

Transcription. The confidentiality of the interviewees needs to be protected and there is also the question of whether a transcribed text is loyal to the interviewee's oral statements.

Analysis. Ethical issues in analysis involve the question of how penetratingly the interviews can be analyzed and of whether the subjects should have a say in how their statements are interpreted.

Verification. It is the researcher's ethical responsibility to report knowledge that is as secured and verified as possible. This involves the issue of how critically an interviewee may be questioned.

Reporting. There is again the issue of confidentiality when reporting private interviews in public, and of the consequences of the published report for the interviewees and for the groups they belong to.

Some of the ethical concerns that can arise throughout seven stages of an interview inquiry are depicted in Box 4.1. These stages will be treated in more detail in Chapter 6 on designing an investigation.

Ethical issues like those presented above should be taken into consideration when preparing an *ethical protocol* for an interview study. Within some fields, such as in the health sciences, it is mandatory to submit an interview project to an ethical review board before the investigation may be undertaken. The researcher is thereby required to think through in advance value issues and ethical dilemmas that may arise during an interview project and perhaps also encouraged to consult experienced members of the research community. Even when it is not a formal requirement, it may be of value when planning an interview inquiry to draft an ethical protocol treating ethical issues that can be anticipated in an investigation. With a foreknowledge of the moral issues that typically arise at the different stages of an interview investigation, the researcher can make reflected choices while designing a study and be alert to critical and sensitive issues that may turn up during the inquiry. In some countries and disciplines, the mandatory submission of an ethical protocol for a research project to institutional review boards may, however, stifle qualitative research.

BOX 4.2   Institutional Review Boards

In the United States, university researchers are required to submit their research projects with human subjects to an institutional review board (IRB). This also holds for several other countries. The projects are reviewed by a university board, according to ethical guidelines for research with human subjects. The guidelines were originally developed for experimentation in biomedical research, where cases of gross misconduct had been uncovered.

When these medical ethical guidelines are extrapolated to the social sciences, several problems arise. While ethical requirements such as fully informed consent are highly pertinent in high-risk medical experiments, they are often unneccesary and unfeasible in low-risk field studies and interviews. The ethical requirements worked out for highly formalized experiments and tests are also, to a certain extent, incongruent with the exploratory, interactive, and interpretative nature of qualitative interviewing, field research, and participatory action research. As expressed by a group of British researchers: "There are inherent tensions in qualitative research that is characterized by fluidity and inductive uncertainty, and ethical guidelines that are static and increasingly formalized" (Birch et al., 2002, p. 2).

The difficulty of specifying in advance the topics of interview studies, which are often exploratory, as well as of describing in advance the specific questions to be posed in a flexible nonstandardized interview, constitutes a potential problem with the ethical review boards. Some boards may want to approve every interview question in advance, which may be feasible for the predetermined questions in a survey interview, but open research interviews involve on-the-spot decisions about following up unanticipated leads from the subjects with questions that cannot be determined in advance. Since many ethical review boards have few specialists in qualitative research as members, their decisions may not be adequate for the complex and open objectives of explorative qualitative projects.

The institutional review boards and their ethical guidelines for human subjects research have been criticized by Lincoln (2005) for serving a new methodological conservatism, constraining participatory qualitative research. She argues further that with the reconfigured relationships of qualitative research as cooperative, mutual, democratic, and open-ended, key issues of common ethical guidelines become nonissues. Parker (2005) has likewise criticized ethics committees in the United Kingdom for favoring quantitative over qualitative approaches, indirectly preventing new forms of research that have not been described in the code, and for being bureaucratic in their use of checklists, often with the result that researchers spend their time trying to get through the review process instead of engaging in serious thought about ethics.

We would like to add that in some countries, such as Denmark, the acceptance of ethical review boards is demanded in health research, whereas for the less life-threatening social research, scholars are themselves considered competent to judge the ethical consequences of their practices. In cases of doubt, they consult more experienced colleagues, and novice researchers will have their advisors to consult, and in complex cases they may consult the national ethical board of their disicipline.

See "Interpreting the Common Rule for the Protection of Human Subjects for Behavioral and Social Science Research" (National Science Foundation, Office of Budget, Finance and Award Management, 2006: http://www.nsf .gov/bfa/dias/policy/hsfaqs.jsp#s) for more specific information about the work and requirements of the institutional review boards in the United States. For critiques of how these relate to qualitative research, see the special issue of *Qualitative Inquiry* edited by Cannella and Lincoln (2007), on "predatory vs. dialogic ethics."

In Box 4.2, we have depicted the institutional review boards evaluating the ethical aspects of research projects and gone through some of the critical points that have been raised by qualitative researchers concerning these boards. There is a danger that it turns out to be so cumbersome and time consuming to translate open exploratory qualitative projects into the formal language of ethical guidelines for experimental research that researchers interested in doing and teaching qualitative research are tempted to give up and seek other research approaches. We further note a parallel of the transfer of ethical codes derived from medical experimental research to social research and the evidence-based research criteria developed for biomedical research relocated to the social sciences (see Box 3.1). In both cases there takes place a bureaucratization of social science research in line with the positivist assumption of a common scientific method regardless of the subject matter investigated. We now turn from the institutional power aspects of research ethics to a discussion of principal ethical positions.

ETHICAL POSITIONS: RULES AND PROCEDURES OR PERSONAL VIRTUES?

In spite of the bureaucracies of ethical review boards and in spite of a conspicuous gap between abstract ethical principles and concrete moral practice, professional ethical codes can, we believe, serve as contexts for reflection on the specific ethical decisions throughout an interview inquiry. Philosophical ethical theories provide frames for more extended ethical reflection; key positions are here a Kantian ethics of duty, a utilitarian ethic of consequences, and Aristotle's virtue ethics (see Eisner & Peshkin, 1990; Kimmel, 1988; and for a broader introduction to moral philosophy see MacIntyre, 2006). Such conceptual contexts seldom provide definite answers to the normative choices to be made during a research project; they are more like texts to be interpreted with respect to their relevance for specific situations.

Until the latter half of the 20th century, there were two dominant moral philosophies in the English-speaking world: Kantian deontology and consequentialism (notably utilitarianism). Kantians (modern Kantians include Jürgen Habermas and John Rawls) have tried to devise a universal procedure that will generate just moral rules and principles, binding to all rational creatures. Utilitarians such as Hume and Bentham (R. M. Hare and Peter Singer are modern representatives) argue that the relevant moral procedure is a kind of universal calculus with which to compute the greatest sum of happiness for all sentient creatures. In recent years, however, a number of philosophers and social scientists have questioned these *procedural* approaches to ethics. The chief problem with these approaches, which are sometimes also called ethics of rules, is that no rule, principle, or procedure can be self-interpreting (Jonsen & Toulmin, 1988). Even if we succeed in formulating a general rule from our procedures that all can agree upon, we still need to know when and how to apply the rule, as we shall see below.

A third position that has become increasingly influential in moral philosophy since the 1960s, Aristotle's virtue ethics, does not primarily aim to formulate a universal *theory* about morality, but rather has the *practical* aim of making us good persons. With virtue ethics, one can de-emphasize the craving for universal ethical procedures and look instead at concretely existing moral practices and personal capabilities. According to Aristotle, what is needed in order to be practically wise in moral matters is not primarily scientific knowledge or ethical principles, but *phronesis,* or prudence, which, as he said, "deals with the ultimate particular thing, which cannot be apprehended by Scientific Knowledge, but only by perception" (Aristotle, 1994, p. 351). *Phronesis* is "the ability to appraise and act upon particular situations in a way that is conducive to the creditable overall conduct of life." (Lovibond, 1995, p. 101).

Phronesis or practical wisdom can be said to involve what we call the skill of "thick ethical description," the ability to see and describe events in their value-laden contexts, and judge accordingly (Brinkmann & Kvale, 2005). The notion of "thick description" was originally proposed by the philosopher Gilbert Ryle and later taken up by the anthropologist Clifford Geertz. We believe that qualitative researchers should primarily cultivate their ability to perceive and judge "thickly" (i.e., using their practical wisdom) in order to be ethically proficient, rather than mechanically follow universal rules. They should engage in contextualized methods of reasoning rather than calculating from abstract and universal principles. Examples and case studies may serve as aids for the transition from general principles to specific practices. The ethical skills embodied in local professional communities further represent an important extension of the written ethical principles, rules, and examples.

However, there is no need to completely abandon rules and principles, for moral rules are still useful as rules of thumb, as Nussbaum (1986) has argued. Moral rules should not in themselves be seen as authoritative, but they are "descriptive summaries of good judgments . . . valid only insofar as they transmit in economical form the normative force of good concrete decisions of the wise person" (Nussbaum, 1986, p. 299). We believe Nussbaum's description of the wise perceiving agent can be read not just as the description of an ethical ideal, but just as much as a description of an ideal qualitative researcher, who knows her subject matter well and can engage in contextualized moral reasoning:

> Being responsibly committed to the world of value before her, the perceiving agent can be counted on to investigate and scrutinize the nature of each item and each situation, to respond to what is there before her with full sensitivity and imaginative vigor, not to fall short of what is there to be seen and felt because of evasiveness, scientific abstractness, or a love of simplification. The Aristotelian agent is a person whom we could trust to describe a complex situation with full concreteness of detail and emotional shading, missing nothing of practical relevance. (Nussbaum, 1990, p. 84)

ETHICAL GUIDELINES

Which ethical questions can we then, as a rule of thumb, say should in most cases be raised at the beginning of an interview study? In what follows, we address four of the fields that are traditionally discussed in ethical guidelines for researchers: informed consent, confidentiality, consequences, and the role of the researcher (see American Psychological Association, 2002; Eisner & Peshkin, 1990; *Guidelines for the Protection of Human Subjects,* 1992).

BOX 4.3   Ethical Questions at the Start of an Interview Study

What are the *beneficial* consequences of the study?

How can the study contribute to enhancing the situation of the participating subjects? Of the group they represent? Of the human condition?

How can the *informed consent* of the participating subjects be obtained?

How much information about the study needs to be given in advance, and what can wait until a debriefing after the interviews?

Who should give the consent—the subjects or their superiors?

How can the *confidentiality* of the interview subjects be protected?

How important is it that the subjects remain anonymous?

How can the identity of the subjects be disguised?

Who will have access to the interviews?

Can legal problems concerning protection of the subjects' anonymity be expected?

What are the *consequences* of the study for the participating subjects?

Will any potential harm to the subjects be outweighed by potential benefits?

Will the interviews approximate therapeutic relationships, and if so, what precautions can be taken?

When publishing the study, what consequences may be anticipated for the subjects and for the groups they represent?

How will the *researcher's role* affect the study?

How can a researcher avoid co-option from the funding of a project or overidentification with his or her subjects, thereby losing critical perspective on the knowledge produced?

In Box 4.3 we attempt through some simple questions to give an overview of some of the questions that may arise in relation to the four key areas of common ethical guidelines, to be addressed in more detail below. Rather than seeing these as questions that can be settled once and for all in advance of the research project, we conceptualize them as *fields of uncertainty* (i.e., problem areas that should continually be addressed and reflected upon throughout an interview inquiry). Rather than attempting to "solve" the problems of consent, confidentiality, and so on once and for all, qualitative research interviewers work in an area where it often is more important to remain *open* to the dilemmas, ambivalences, and conflicts that are bound to arise throughout the research process. This demands going beyond the ethical guidelines and principles and focusing more on the ethical capabilities of researchers.

A brief example from Plato demonstrates our main point that there are no ethical rules or principles that are self-applying and self-interpreting—rules must always be understood contextually. In a dialogue from *The Republic*, Socrates is in a conversation with Cephalus, who believes that justice—here, "doing right"—can be stated in universal rules, such as "tell the truth" and "return borrowed items":

> 'That's fair enough, Cephalus,' I [Socrates] said. 'But are we really to say that doing right consists simply and solely in truthfulness and returning anything we have borrowed? Are those not actions that can be sometimes right and sometimes wrong? For instance, if one borrowed a weapon from a friend who subsequently went out of his mind and then asked for it back, surely it would be generally agreed that one ought not to return it, and that it would not be right to do so, not to consent to tell the strict truth to a madman?'
>
> 'That is true,' he replied.
>
> 'Well then,' I [Socrates] said, 'telling the truth and returning what we have borrowed is not the definition of doing right.' (Plato, 1987, pp. 65–66)

Incidentally, this conversation is an example of what we shall address as the confrontational interview in Chapter 8, and it has the aim of producing knowledge in the sense of *episteme,* as discussed in Chapter 3 (see also Brinkmann, 2007a). The knowledge that goes forth can be summarized in the statement that moral rules, guidelines, and principles should not be applied mechanically, for there are always situational factors that determine when and how they are morally relevant. Such factors cannot themselves be stated in explicit rules, for we end in an infinite regress if we try to state the rules for when to apply the rules. Thus, we recommend that ethical guidelines should be reconfigured pragmatically as tools to think with in fields of uncertainty, rather than being seen as the final moral authority that ignores real-life ambiguities and uncertainties. We shall now turn to fields of ethical uncertainty in interview research normally addressed by ethical guidelines.

Informed Consent

Informed consent entails informing the research participants about the overall purpose of the investigation and the main features of the design, as well as of any possible risks and benefits from participation in the research project. Informed consent further involves obtaining the voluntary participation of the people involved, and informing them of their right to withdraw from the study at any time.

Through briefing and debriefing, the participants should be informed about the purpose and the procedures of the research project. This should include information about confidentiality and who will have access to the interview or other material, the researcher's right to publish the whole interview or parts of it, and the participant's possible access to the transcription and the analysis of the qualitative data. In most cases, such issues may not matter much to the participants. If, however, it is likely that the investigation will treat or instigate issues of conflict, particularly within institutional settings, a written agreement may serve as protection for both the participants and the researcher. When it comes to later use of the research material, it may be preferable to obtain a written agreement, signed by both researcher and participant, with the informed consent of an interviewee to participate in the study and allow future use of the material (see Yow, 1994, for examples of letters of agreement with participants).

Issues about *who should give the consent* may arise when doing research in institutions where a superior's consent to a study may imply a more or less subtle pressure on his or her subordinates to participate. With schoolchildren, the question comes up about who should give the consent—the children themselves, their parents, the teacher, the school superintendent, or the school board?

Informed consent also involves the question of *how much information should be given and when.* Full information about design and purpose counteracts deception of the participants. Providing information about a study involves a careful balance between giving too much detailed information and leaving out aspects of the design that may be significant to the participants. In some interview investigations, such as those using funnel-shaped questioning techniques that gradually narrow down on the subject matter, the specific purposes of a study are initially withheld in order to obtain the interviewees' spontaneous views on a topic and to avoid leading them to specific answers. In these cases, full information should be given in a debriefing after the interview.

In many case studies, interviews are often used as one technique alongside others. For example, if one is doing ethnographic fieldwork, interviews can merely be brief moments in a much longer time period of participant observation. In such forms of qualitative inquiry, it can sometimes be difficult to determine when the more informal interaction ends and an interview begins. This challenges the researcher to consider the ethical implications of informing the respondents of how and what aspects of their interaction will be used in the further research process.

Informed consent as an ethical field of uncertainty thematizes the conflict between a complete disclosure of the rationale of the research project beforehand (thereby rendering much qualitative interview research impossible) and withholding information from the participants, which may sometimes result in knowledge that can improve the condition of the larger community. There is further the issue of how informed consent can be handled in exploratory qualitative studies where the investigators themselves have little advance knowledge of how the interviews and observations will proceed.

Confidentiality

Confidentiality in research implies that private data identifying the participants will not be disclosed. If a study will publish information that is potentially recognizable to others, the participants should agree to the release of identifiable information. The principle of the research participants' right to privacy is not without ethical and scientific dilemmas.

There is thus a concern about *what information should be available to whom.* Should, for example, interviews with children be available to their parents and teachers? In studies where several parties are involved (e.g., in the case of individual interviews within organizations or with married or divorced couples), it should be made clear before the interviewing who will later have access to the interviews. Protecting confidentiality can in extreme cases raise serious legal problems, such as in cases when a researcher—through the promise of confidentiality and the trust of the relationship—has obtained knowledge of mistreatment, malpractice, child abuse, the use of drugs, or other criminal behaviors by either the participants or others. In the United States, researchers may in advance obtain a certificate of confidentiality from the federal government, protecting against disclosure of the identity of their participants (see *Guidelines,* 1992, p. 6).

Qualitative methods such as interviews involve different ethical issues than questionnaire surveys, where confidentiality is assured by the computed averages of survey responses. In a qualitative interview study, where participants' statements from a private interview setting may appear in public reports, precautions need to be taken to protect the participants' privacy. There may here be an intrinsic conflict between the ethical demand for confidentiality and the basic principles of scientific research, such as providing the necessary specific information for inter-subjective control and for repeating a study.

Confidentiality as an ethical field of uncertainty relates to the issue that, on the one hand, anonymity can protect the participants and is thus an ethical demand, but, on the other hand, it can serve as an alibi for the researchers, potentially enabling them to interpret the participants' statements without being gainsaid. Anonymity can protect the participants, but it can also deny them "the very voice in the research that might originally have been claimed as its aim" (Parker, 2005, p. 17). We should also note that in some cases interviewees, who have spent their time and provided valuable information to the researcher, might wish, as in a journalistic interview, to be credited with their full name.

Consequences

The consequences of a qualitative study need to be addressed with respect to possible harm to the participants as well as to the benefits expected from their participation in the study. The ethical principle of *beneficence* means that the risk of harm to a participant should be the least possible (*Guidelines,* 1992, p. 15). From a utilitarian ethical perspective, the sum of potential benefits to a participant and the importance of the knowledge gained should outweigh the risk of harm to the participant and thus warrant a decision to carry out the study. This involves a researcher's responsibility to reflect on the possible consequences not only for the persons taking part in the study, but also for the larger group they represent.

The researcher should be aware that the openness and intimacy of much qualitative research may be seductive and can lead participants to disclose information they may later regret having shared. A research interviewer's ability to listen attentively may also, in some cases, lead to quasi-therapeutic relationships, for which most qualitative researchers are not trained. In particular, long and repeated interviews on personal topics may lead to quasi-therapeutic relationships. The personal closeness of the research relationship puts continual and strong demands on the tact of the researcher regarding how far to go in his or her inquiries.

Anticipating potential ethical transgressions also requires a thorough knowledge of the field of inquiry. Some interview researchers, who had been oblivious to the significant differences between oral and written language, may thus have recollections of having hurt the dignity of their interviewees when they sent the verbatim transcriptions of interviews to the interviewees to have them check and validate the statements. The field of uncertainty that opens up when we consider the consequences of qualitative research is perhaps the most complex one, because it is often unpredictable. If a conversation between a researcher and a participant suddenly takes a turn and touches sensitive issues that obviously move the participant, how should the researcher react? Should the researcher pursue these issues in a therapeutic vein in order to help the participant (and perhaps obtain important knowledge as a "side effect"), but with the risk of ethically transgressing the participant's intimate sphere, or should the researcher refrain from anything resembling therapeutic intervention with the ethical risk of appearing cold and aloof?

The Role of the Researcher

The role of the researcher as a person, of the researcher's integrity, is critical to the quality of the scientific knowledge and the soundness of ethical decisions in qualitative inquiry. Morally responsible research behavior is more than abstract ethical knowledge and cognitive choices; it involves the moral integrity of the researcher, his or her sensitivity and commitment to moral issues and action. In interviewing, the importance of the researcher's integrity is magnified because the interviewer him- or herself is the main instrument for obtaining knowledge. Being familiar with value issues, ethical guidelines, and ethical theories may help the researcher to make choices that weigh ethical versus scientific concerns in a study. In the end, however, the integrity of the researcher—his or her knowledge, experience, honesty, and fairness—is the decisive factor.

To the ethical requirements of the researcher also belongs a strict adherence to the *scientific quality* of the knowledge published. This involves publishing findings that are as accurate and representative of the field of inquiry as possible. The results reported should be checked and validated as fully as possible, and with an effort toward a transparency of the procedures by which the conclusions have been arrived at. It can often be difficult to determine whether such requirements are ethical, concerned with good conduct in a moral sense, or epistemic, concerned with actions that lead to the production of significant knowledge (Brinkmann, 2007b).

The *independence of research* can be co-opted from "above" as well as "below," by those funding a project, as well as by its participants. Ties to either group may lead the researcher to ignore some findings and emphasize others to the detriment of as full and unbiased an investigation of the phenomena as possible. Qualitative interview research is interactive research; through close interpersonal interactions with their participants, interviewers may be particularly prone to co-optation by them. The researchers may so closely identify with their participants that they do not maintain a professional distance, but instead report and interpret everything from their participants' perspectives, "going native" in anthropological language.

The field of uncertainty that is disclosed when reflecting on the role of the researcher can involve a tension between a professional distance and a personal friendship. Thus in the context of a feminist, caring, committed ethic, the qualitative research interviewer has been conceived of as a friend, a warm and caring researcher. This early conception of the researcher as a caring friend was subsequently criticized from a feminist standpoint (Burman, 1997). Duncombe and Jessop (2002) argue that an interviewer's show of intimacy and empathy may involve a faking of friendship and commodification of rapport, sanitized of any concern with broader ethical issues. When interviewers are under pressure to deliver results, whether to a commercial employer or to their own research publications, their show of empathy may become a means to circumvent the participant's informed consent and persuade interviewees to disclose experiences and emotions that they later decide they may have preferred to keep to themselves or even "not know." With an expression from a therapist-researcher (Fog, 2004), an experienced interviewers' knowledge of how to create rapport and get through a participant's defenses may serve as a "Trojan horse" to get inside areas of a person's life where they were not invited. The use of such indirect techniques, which are ethically legitimate within the mutual interest of therapeutic relations, become ethically questionable when applied to research and commercial purposes.

Psychotherapeutic researchers can in their interviews go further than academic interviewers with regard to some of the potential consequences for those they interview. They may thus deliberately stimulate strong emotional bonds, circumvent their defenses, provoke anxiety, attempt to revive painful memories, put forth critical interpretations of what the patients say and do, and stimulate changes in the patients' behavior, which are all interventions implied in an implicit contract with the patient to help him or her to improve. We may venture that the richness of new knowledge produced by psychoanalytic interviews may in part be due to the psychotherapists' "ethical license" to address human phenomena in ways which normally are out of bonds for academic interview researchers.

The four fields of uncertainty, concerning informed consent, confidentiality, consequences, and the researcher's role, can be used as a framework when preparing an ethical protocol for a qualitative study, and they can be used as ethical reminders of what to look for in practice when doing interview research. Ethical issues typically arise in interview research because of the asymmetrical power relation between interviewer and respondent, where, as we emphasized in Chapter 2, researchers are usually positioned as the relatively more powerful side. Sometimes, the practice of qualitative research is portrayed as inherently dominance free, based on trust and empathy and a free exchange of viewpoints, but such assertions overlook important dimensions of qualitative research involving power (Brinkmann & Kvale, 2005; Kvale, 2006). Some interview researchers have portrayed qualitative inquiry as inherently ethical, or at least more ethical than quantitative research. This qualitative progressiveness myth has been baptized "qualitative ethicism" by Hammersley (1999). It is the tendency to see research almost exclusively in ethical terms, as if the rationale of research was to achieve ethical goals and ideals with the further caveat that qualitative research uniquely embodies such ideals. But, as testified to by the four fields of uncertainty, qualitative interview research is laden with just as many complex ethical issues as other forms of research that involve human participants. Our point is not that qualitative research is particularly ethically suspect, but nor is it particularly ethically good in itself. Here, it could be relevant to follow Foucault's ethical advice: "My point is not that everything is bad, but that everything is dangerous, which is not exactly the same as bad. If everything is dangerous, then we always have something to do." (1997, p. 256). Regarding ethics, researchers always have something to do, and this is also true for those in the field of qualitative research interviewing.

LEARNING ETHICAL RESEARCH BEHAVIOR

According to the practical outlook on morality originally proposed by Aristotle, the task of ethics is not to provide an abstract theory of the good, but rather to make us good (Aristotle, 1994). From the Aristotelian viewpoint that informs this chapter, our moral reality is a practical reality where truthfulness is more important than absolute truth, and where *phronesis* or practical wisdom—the skill of clear perception and judgment—becomes more important than theoretical understanding and the ability to use abstract procedures. The lesson to learn here is that by describing the qualitative world adequately, by getting close enough to phenomena, by being objective concerning particular situations, we will be lent a hand in knowing what to do that goes beyond formal ethical guidelines and the abstract principles of ethical theories. The reduced role of theoretical criteria for determining moral action does not mean that we have no criteria at all. Rather than demanding theoretical proof in the moral realm, we simply ought to try to act well in accordance with our practical wisdom (Levine, 1998). Stephen Toulmin, a philosopher who has also served on ethical commissions, has observed that commissioners can often reach consensus concerning particular cases, but not on the formal principles and abstract theories that might justify their concrete decisions (Toulmin, 1981). As pragmatists like Toulmin like to emphasize, it does seem possible to engage in morally reasonable and responsible action without complete agreement on universal ethical theory. For pragmatists, moral conduct is more akin to a skilled craft than to the logic of mathematical reasoning.

With the emphasis on the ethical capabilities and virtues of the qualitative researcher, the learning of ethical research behavior becomes a key issue. Again, a relevant source here is Aristotle's virtue ethics, which has also inspired the current movement in social ethics known as *communitarianism* (see Mulhall & Swift, 1996). Communitarians reject "the liberal self," the autonomous, isolated chooser presupposed in rational-choice theory and contractualism. Antithetical to communitarianism is the idea of independent, autonomous agents who enter freely and knowingly into contractual relations that is the basis for the proliferation of ethical codes and committees today. In communitarianism, the self is conceived as constituted by communal attachments within communities and traditions. A background of communities and traditions is needed in order to learn ethical research behavior, and we suggest two approaches to learning ethical behavior here. The first is the skill model of Dreyfus and Dreyfus (1990). According to their phenomenological account, the development of ethical capabilities can be described as a five-step ladder, starting with explicit rules and reasoning, which, with increasing experience and expertise, recede into the background of skill and habit, where the highest form of ethical comportment consists of being able to stay involved and to refine one's intuitions. Moral consciousness then consists of unreflective responses to interpersonal situations, which in cases of disagreement may be attempted to be solved in a dialogue.

The second approach to learning ethical research behavior is through an increased mastering of the art of thick ethical description in relation to contexts, narratives, examples, and communities. These represent four ways of learning to "thicken" events to help us act morally:

Contextualize. We thicken events by describing them in their context. In a court of law, for example, the question whether somebody did something intentionally is decided not by citing theories or general rules, but by describing the context of the act. Thick description situates an event in a context, and the experienced ethical reasoner knows which features of a context are relevant in order to judge adequately. The skilled qualitative researcher understands the peculiar features of the research context, and how this context generates specific ethical issues to be addressed, which parallels the contextual features of interview knowledge that were addressed in the previous chapter.

Narrativize. Those thick descriptions that incorporate a temporal dimension in a storied form are called *narratives,* and "narratives can carry moral meaning without relying upon general principles" (Levine, 1998, p. 5). If we manage to pull together a convincing narrative that situates an event temporally and socially, then we rarely need to engage in further moral deliberation about what to do. Looking at a situation in a "snapshot," outside its temporal and social narrative context, will on the other hand make it hard to judge and act morally. If one is not provided with the kind of information necessary to narrativize—for example, if the researcher has never met the participant before and does not know her larger life story—then it is ethically wise to be lenient about one's interpretations and generalizations, and refrain from anything resembling therapeutic intervention.

Focus on the Particular Example. Within a virtue conception of ethics, Løvlie (1993) has attempted to overcome the opposition of explicit rules versus tacit skills by the introduction of examples. These may be in the form of parables, allegories, myths, sagas, morality plays, case histories, and personal examples, all in the form of thick ethical descriptions. The qualitative researcher should know about exemplars of ethically justifiable and also ethically questionable research, in order to evaluate his or her practice and learn to recognize ethical issues. Generalizations, as found in formal ethical guidelines, should not blind us to the crucial particularities encountered in a specific research situation.

As qualitative researchers are involved in actual issues with particular people at particular places and times, they need to master an understanding of these concrete particulars in order to be morally skillful.

Consult the Community of Practice. The learning of ethical research behavior is a matter of being initiated into the mores of the local professional culture. Our emphasis above on a shift from an ethics of rules to the ethical capabilities of the qualitative researcher should not imply an "ethical overburdening" of the individual researcher. Qualitative research is rarely practiced by single researchers who confront their participants as isolated individuals. A researcher is usually part of a research community, and is normally accountable not only to the participants but also to peers, superiors, students, his or her institution, and the discipline at large. When confronted with difficult ethical issues, it is often wise to consult the research community, and if one wants to improve the skills of ethical perception, judgment, and reasoning, one needs to receive feedback from others. We can only learn by being corrected, and this presupposes the existence of a community with sufficiently shared values and some agreement concerning what behaviors stand in need of being corrected and when. Learning qualitative inquiry in a research community where ethical and scientific values are part of the daily practices may foster integrated research behavior in an ethical sense.

****

In this chapter, we have argued that learning ethical principles is not sufficient to become an ethically responsible qualitative interview researcher. We have pointed to thick ethical description as an approach to learning ethical behavior in qualitative research. Learning to describe particulars thickly does not just involve learning rules, but learning from cases and observing those who are more experienced. It is about learning to *see* and *judge* rather than learning to universalize or calculate. Interestingly, the art of thick ethical description is similar to what the good (in a nonmoral sense) qualitative researcher should master in order to produce new, insightful knowledge about the human condition. Ethical proficiency, in this perspective, involves dimensions of skill and craftsmanship, knowing the good practices of the trade, and in the next chapter we go more directly into a discussion of whether qualitative interviewing is most fruitfully conceived as a method or as a craft.

FIVE

LEARNING THE CRAFT OF QUALITATIVE RESEARCH INTERVIEWING

———◆◆◆———

W e conclude this part on conceptualizing qualitative research interviewing by discussing how to conceive of the practice of research
interviewing. We first address interviewing as a rule-governed method versus
the exercise of personal skills and judgment. We then go on to outline research
interviewing as a craft, after which we turn to forms of learning the interview
craft such as research apprenticeships and research practica, which take into
account the acquisition of the tacit and context-bound aspects of skills.

Exercise for Starting to Learn Interviewing as a Craft
Readers who want to learn interviewing skills the way a craft is learned
should stop reading the book now and perform the following task:
*Obtain about three sound-recorded research interviews, spend a week
transcribing them, and reflect on the processes and problems of transcribing and interviewing.*
After this hands-on experience, you may continue by reading the practical
"how to" suggestions in this book in Chapters 6 to 16. Most likely, you will then
discover that you have already experienced through your own practice much of
what is written on the following pages. Some of the expected learning outcomes
are mentioned in Box 5.2, "Learning Interviewing by Transcribing Interviews."

RESEARCH INTERVIEWING:
METHOD OR PERSONAL SKILLS

A priority of method over skills has prevailed in modern social science. In qualitative interview research today there is an emphasis on interviewing as a method and, in particular, on methods for analyzing the transcribed interviews. *Method* may be understood in a broad sense, following the original Greek meaning of the term as "the way to a goal." In the bureaucratic and positivist approaches to the social sciences, however, method has become confined to a mechanical following of rules. This definition is an example: "A method is a set of rules, which can be used in a mechanical way to realize a given aim. The mechanical element is important: a method shall not presuppose judgment, artistic or other creative abilities" (Elster, 1980, p. 295).

A *survey interview* may live up to this formal definition of method. The interviewer here follows standard rules with a minimum of personal judgment and poses questions in a predetermined sequence and wording. The data produced in the survey interview should be exactly reproducible by other interviewers. Expertise and craftsmanship in survey research is manifested in the construction of questions and in the mastery of statistical techniques for data analysis. The very questioning that produces the survey data is, however, a semiskilled activity; it does not require interviewer expertise in various questioning techniques or knowledge of the subject matter investigated, and it can be learned in a matter of hours.

In a *qualitative research interview,* knowledge is produced socially in the interaction of interviewer and interviewee. The very production of data in the qualitative interview goes beyond a mechanical following of rules and rests upon the interviewers' skills and situated personal judgment in the posing of questions. Knowledge of the topic of the interview is in particular required for the art of posing second questions when following up the interviewee's answers. The quality of the data produced in a qualitative interview depends on the quality of the interviewer's skills and subject matter knowledge. Extensive training is required to become a highly qualified interviewer. With a conception of method as a mechanical following of a set of rules, the production of knowledge through the personal interaction between interviewer and interviewee is clearly not a legitimate scientific method.

The antithesis to the conception of research as a craft is the methodological positivists' idea that strict adherence to certain prespecified rules is a truth guarantee in science. As long as one follows the methodical rules, it is claimed, personal skills and capabilities are irrelevant in the production of knowledge, as in the impersonal and mechanical definition of method above. Philosophers have long questioned this idea: In *Truth and Method,* Gadamer (1975) argued that knowledge in the humanities cannot be reduced to a method, for we can only know the social and historical world through understanding and interpretation, which ultimately rest on pre-understandings and pre-judices that cannot be codified into methodological rules. For the natural sciences, Feyerabend (1975) argued in *Against Method* that if the canon of significant scientists had followed the methodical prescriptions of the philosophy of science, key discoveries could never have been made—such discoveries are often a result of breaking rather than following the standard rules of research.

An antimethod approach to social research, with an emphasis on the person of the researcher as the very research instrument, has been expressed by several social researchers who have contributed with substantial new knowledge to their fields, for example by the anthropologist Jean Lave in this interview sequence:

SK: Is there an anthropological method? If yes, what is an anthropological method?

JL: I think it is complete nonsense to say that we have a method. First of all I don't think that anyone should have *a* method. But in the sense that there are "instruments" that characterise the "methods" of different disciplines—sociological surveys, questionnaire methods, in psychology various kinds of tests and also experiments—there are some very specific technical ways of inquiring into the world. Anthropologists refuse to take those as proper ways to study human beings. I think the most general view is that the only instrument that is sufficiently complex to comprehend and learn about human existence is another human. And so what you use is your own life and your own experience in the world. (Lave & Kvale 1995, p. 220)

Bourdieu, who was rather critical of the multitude of methodological writings in the social sciences, likewise emphasized the role of the researcher as a person, of his or her respect and sensitivity to social relations. In *The Weight of the World* (1999) he, somewhat reluctantly, added a chapter about his own understanding of interviewing:

> I do not believe that it is useful to turn to the innumerable so-called "methodological" writings on interview techniques. . . . At any rate it does not seem to me that they do justice to what has always been done—and known—by researchers, who have the most respect for their object and who are the most attentive to the almost infinitely subtle strategies that social agents deploy in the ordinary conduct of their lives. . . . the adequate scientific expression of this practice is to be found neither in the prescriptions of a methodology more often scientistic than scientific, nor in the antiscientific caveats of the advocates of mystic union. (p. 607)

The role of the interviewer as a person also is decisive in the psychotherapeutic interview. Sullivan (1954) has analyzed the therapeutic interview as an interpersonal situation, where the relevant data are constituted by the interaction itself, in the specific situation created between interviewer and interviewee. Sullivan emphasized the subjective moment in obtaining knowledge—in participant observation it is the interviewer as a person who is the method, the instrument. We conclude that while survey interviewing may be adequately described as a method in a strict sense, qualitative research interviewing is closer to a craft. The personal skills and respect needed to practice qualitative research interviewing excellently cannot be reduced to methodological rules. In the following section we attempt to describe the craft of research interviewing.

THE CRAFT OF RESEARCH INTERVIEWING

An emphasis on the crucial role of the interviewer as a person does not imply neglect of techniques and knowledge. When the person of the researcher becomes the main research instrument, the competence and craftsmanship—the skills, sensitivity, and knowledge—of the researcher become essential for the quality of the knowledge produced. A conception of interviewing as a craft may lead interview research clear of the Scylla of methodological objectivism and the Charybdis of subjectivist relativism.

We should note that the conception of research interviewing as a craft does not involve a mere reliance on personal intuition or an anything goes approach. C. Wright Mills (1959/2000) has thus depicted social research in general as "intellectual craftsmanship." A mastery of the relevant methods and theories is thus important for the craft of research interviewing, but should not become an autonomous fetish of scientific inquiry. According to Mills, methods and theories "are like the language of the country you live in; it is nothing to brag about that you can speak it, but it is a disgrace and an inconvenience if you cannot" (1959/2000, p. 121). He wrote,

> Be a good craftsman: Avoid any rigid set of procedures. Above all, seek to develop and to use the sociological imagination. Avoid the fetishism of method and technique. Urge the rehabilitation of the unpretentious intellectual craftsman, and try to become such a craftsman yourself. Let every man be his own methodologist; let every man be his own theorist; let theory and method again become part of the practice of a craft. (Mills, 1959/2000, p. 224)

Before pursuing the implications of conceiving research interviewing as a craft, we shall briefly relate it to three other conceptions of research interviewing.

BOX 5.1   Four Conceptions of Research Interviewing

Interviewing as Semiskilled Labor

The interview process is predetermined and routinized in the form of standardized operating procedures to be followed in identical ways by all interviewers in a study. Interviewers are expected to implement the interview guide in the prescribed manner and adhere to specific procedures under close supervision. Survey interviewers can learn to mechanically follow the rules of a given method in a matter of hours.

Interviewing as a Skilled Craft

Interviewing is seen as a craft, requiring a repertoire of specialized skills and the exertion of personal judgment. The interviewer possesses a "toolbox" with special techniques and rules of thumb for their application and has a knowledge of the interview topic. The interviewer who performs his or her craft without detailed supervision may require an extended practice to become an able interviewer.

Interviewing as Requiring Professional Expertise

Interviewing not only requires a repertoire of specialized skills, but also demands the exercise of qualified judgment in their application, as well as

(Continued)

(Continued)

a situated judgment of the consequences of the practical decisions in a broader social context (see the description of phronesis in Chapter 4). To exercise independent and sound professional judgment, the interviewer is expected to master theoretical knowledge of interviewing and of the interview topic. The understanding of interviewing put forth in this chapter encompasses both a craft and a professional conception of research interviewing; see Mills's (1959/2000) characterization of research as "intellectual craftsmanship."

Interviewing as an Art

Interviewing as an art involves intuition, creativity, improvisation, and breaking the rules. The interviewing techniques may be unconventional and novel. Techniques and standards of practice are used in a personal rather than a standardized way. The art interviewer has considerable autonomy in the performance of interviewing. Eisner (1991) has addressed qualitative research in general from an artist's perspective, emphasizing the connoiseur's sensitivitity to qualitative distinctions and the ability to communicate new perspectives.

In Box 5.1 we have outlined four conceptions of research interviewing, inspired by an overview of conceptions of teaching by Darling-Hammond, Wise, and Pease (1986). We regard all four conceptions as legitimate modes of producing knowledge and would like to repeat that when survey interviewing is conceived as a semiskilled form of labor, this concerns only the data-producing process of questioning; survey research as a whole is a demanding process requiring professional expertise. We would also like to point out the paradox that when it comes to the very process of producing data in interviews, the less skill required of the interviewers, the closer the process comes to a mechanical conception of method, and the higher the social scientific status of the interview form. And, correspondingly, the more expertise required in the basic production of interview data, the less the scientific status of the knowledge produced.

By *craftsmanship* we refer to mastery of a form of production, which requires practical skills and personal insight acquired through training and extensive practice. A craft is learned through apprenticeship in a community of practitioners (see Lave & Wenger, 1991), through observation and imitation of examples of best practice, and through feedback on one's own performances. In the traditional crafts there was an emphasis on ethics, on personal responsibility and loyalty to the craft. Sennett has, in his search for values to countervail the culture of the new capitalism, gone back to the commitment of the old crafts: "Craftsmanship broadly understood means the desire to something well for its own sake" (2006, p. 194).

Knowledge of interviewing is embedded less in explicit rules of method than in concrete examples of interviews; the practical skills and personal know-how of research interviewing remain to a large extent content and context bound. The medium of the interview is language, and while important aspects of interviewing can be verbalized, such as the formulation of questions, there remain relational and tacit aspects of interviewing skills, which are difficult to verbalize, not least the art of patient listening.

Good interview research goes beyond formal rules and encompasses more than the technical skills of interviewing to also include personal judgment about which rules and questioning techniques to invoke or not invoke. The proficient craftsman does not focus on the techniques, but on the task and on the material, the object, he or she works with (in Chapter 3, we described this object in terms of seven characteristics). The interviewer who masters his or her trade thinks less of interviewing technique than of the interviewee and the knowledge sought.

Conceptual knowledge of the subject matter of an investigation may serve to create order and meaning when conducting and analyzing interviews. Before going on to the next paragraph, the reader is encouraged to take a moment and try to remember this row of numbers: 1491625364964.

Most readers will require many rehearsals to learn the numbers by rote before being able to recite them correctly. Students of mathematics may, however, remember them perfectly after a single reading. With their mathematical training they will immediately see the numbers as a row of square numbers— one in second, two in second, three in second, and so on—and, once having grasped this principle, they will remember the whole sequence. Knowledge of the subject matter makes it possible to see a principle in the apparent chaotic row of numbers. Within qualitative research it may be harder to find such simple principles to order the many words, but it is still worthwhile to approach interviewing with a linguistic competence in the medium of interviewing— language—and with a knowledge of the structures and functions of language to have one means of creating structure in what often appears a chaos of words.

Professional interview research goes beyond a mastery of interviewing skills and conceptual knowledge of the subject matter of an inquiry. It also involves a theoretical reflection on the production of knowledge in research interviewing, such as addressing the epistemological and ethical issues of interviewing, as well as conceptions of interviewing practice. Conceiving of qualitative research interviewing as a method or as a craft is not a mere play of words, or only an abstract epistemological issue, but has concrete consequences for the practice and the learning of research interviewing. Understanding research interviewing as a method or as a craft involves different logics of practice, and melding the two approaches may lead to a muddled practice and broken expectancies. Thus the methodical requirements of standard predetermined wording and sequences of questions, which are necessary in the method of survey interviewing, will block the force of the qualitative interview craft, which rests on personal competence and judgment in the wording and sequencing of questions. Demands of advance explicit formulations of procedures and questions for a research inquiry, which depend on the skills and know-how of the researcher, marginalize personal intuition, flexibility, and creativity in interview research.

From the above discussion it follows that interviewing as a craft is not some mere prescientific method, which needs to be developed into a formalized rule-governed method to become a legitimate scientific method. The very personal interaction of the interview, and the interpersonal skills required of the interviewer, defy any formalization into impersonal methodic procedures. The development of research interviewing depends on improving the personal skills of the interviewer, to which we turn now.

Mishler has argued that qualitative researchers turn out to resemble craftspeople more than logicians. Their competence depends upon contextual knowledge of the specific methods applicable to a phenomenon of interest: "Skilled research is a craft, and like any craft, it is learned by apprenticeship to competent researchers, by hands-on experience, and by continual practice" (Mishler, 1990, p. 422; see also Hammersley, 2004, and Seale, 2004).

LEARNING THE CRAFT OF RESEARCH INTERVIEWING

As Aristotle noted, there are different kinds of knowledge, which are learned in different ways. Thus mathematics and theory may be learned through books and in school, whereas practical activities are learned in practice:

> We learn an art or craft doing the things that we shall have to do when we have learned it: for instance men become builders by building houses, harpers by playing on the harp. Similarly we become just by doing just acts, temperate by doing temperate acts, brave by doing brave acts. (Aristotle, 1994, p. 73)

Good practice consists of more than merely carrying out practical acts, it also involves a situated judgment of what knowledge and techniques to apply when acting in a given context, and when one is confronted with different goals and values that demand careful choice.

The craft of research interviewing is learned by practicing interviewing, preferably within a community of experienced interviewers. The skills, the knowledge, and the personal judgments necessary for conducting a qualitative interview of a high quality require extensive training. The flexible, content- and context-related skills of interviewing are acquired by doing interviews. Whereas the phrasing of questions can be communicated verbally, other aspects of interview skills, such the intonation of questions, the stretching of pauses, sensitive listening, and the establishing of good rapport in the interview situation rest largely on tacit knowing acquired through practice and by working with experienced interviewers.

At the beginning of this chapter we suggested that readers who wanted to learn interviewing in ways approximating the learning of a craft should stop reading and perform a transcription task. The task is inspired by Tilley's (2003) article "Transcription Work: Learning Through Co-Participation in Research Practices," where she discusses transcription in relation to Lave and Wenger's (1991) analysis of apprenticeship learning in communities of practice.

BOX 5.2   Learning Interviewing by Transcribing Interviews

Obtain about three sound-recorded research interviews, spend a week transcribing them, and reflect on the processes and problems of transcribing and interviewing.

Learning lessons:

- To secure a good quality sound recording
- To clarify inaudible answers during an interview
- To pose clear questions that interview subjects understand

(Continued)

(Continued)

- To listen carefully to what is said and how it is said
- To pay attention to the voice, the pauses, the sighs, and the like, as indications that a topic may be important, and possibly also too sensitive to pursue
- To become sensitive to the possible ethical transgressions when questioning too privately or critically
- To follow up an interview statement with a second question
- To prevent the interview from becoming filled with small talk
- To notice interviewer variations in questioning styles, their advantages and their drawbacks
- To become aware of the differences between oral and written language, and the need of guidelines for translation from oral to written language
- To notice how new interpretations of the meanings may spontaneously arise when working closely with the oral recording

Box 5.2 depicts some potential learning lessons from the transcription task. During a week of transcribing, newcomers may have discovered by themselves much of what is presented in the following chapters on interviewing and transcribing. Starting to learn interviewing by listening to sound recordings will sensitize novice interviewers to the oral medium of the interview craft. Learning interviewing by transcribing interviews promotes discovery learning, where newcomers to the trade, through their own practice, discover techniques and dilemmas of transferring live conversations into written texts. In a safe transcribing situation, the novice may further become aware of the subtleties of interviewing, where he or she will not ruin the knowledge production of an important interview, or ethically transgress the interviewee's intimate sphere in a live interview situation.

As the complex skills of the interview craft can be learned through textbooks and short courses only to a certain level, it becomes necessary to address alternative conceptions and practices for transmitting the craft of interviewing. Here we first depict briefly the skill model of Dreyfus and Dreyfus and research apprenticeship. Then, inspired by the practicum developed by Schön for training reflective practitioners, we suggest practica for learning the interview craft.

Dreyfus and Dreyfus (1986) have proposed a model of skill-learning that encompasses both explicit rule following and intuition. Drawing on examples from driving, playing chess, making medical diagnoses, and nursing, they outline five stages of adults' acquisition of skills: novice, advanced beginner, competence, proficiency, and expertise. What stands out is a progression from the analytic behavior of a detached subject, of a novice learning through instruction of "context-free" elements and combining the facts by "context-free rules," to emotionally involved skillful behavior at the higher level of skills. The expert "sees" or "feels" solutions by relying on intuitive knowledge generalized from extensive case experience. If we follow the Dreyfus model, explicit guidelines for interviewing, such as those put forward in this book, pertain to the beginner levels of interviewer skills, whereas interviewing at an expert level would be attained through practice and would to a large extent be intuitive.

When research interviewing is conceived of as a craft, the ideal way to learn the craft of interviewing is through direct exposure to interview practice in a research apprenticeship. In contrast to the individualist Dreyfus model of skill learning, apprenticeship learning takes place in communities of practice, where even beginners do not need to start by learning explicit rules, but can immediately become immersed in elementary practices of the craft. In an interview apprenticeship, the novices would learn in a community of practice, learning research by participating in research practices, observing and imitating the masters of the trade. Learning in apprenticeship entails not only acquisition of skills, but also incorporation of the values of the trade and its norms of ethical behavior (see Lave and Wenger, 1991, for a general analysis of apprenticeship training; and for apprenticeship training of Nobel laureates in the natural sciences, see Kvale, 1997, and Zuckermann, 1972).

Not every novice interviewer will have access to an apprenticeship in a research group where interviewing is part of the daily practice. Or the researchers may not have time available for instructing novices. One option in the latter case could be, as with apprenticeship learning of the crafts, that the novice "pays for the tuition" by performing the simple tasks necessary for the ongoing research projects, such as transcribing, and learns at the same time. In the analysis stage, there may also be a need for several coders of the transcripts, allowing the learner to work his or her way into the trade by performing the tasks of the trade.

With few authentic research apprenticeships available, the interview craft may be learned in an interview practicum simulating real-world practice:

> A practicum is a setting designed for the task of learning a practice. In a context that approximates a practice world, students learn by doing, although their doing usually falls short of real-world work. They learn by undertaking projects that simulate and simplify practice; or they take on real world projects under close supervision. (Schön, 1991, p. 37)

Schön dismisses the "technical rationality, the school's prevailing epistemology of practice, [which] treats professional competence as the application of privileged knowledge to instrumental problems of practice" (1991, p. xi). On the basis of his case studies and interviews with professionals such as architects, administrators, and psychotherapists, Schön describes a professional knowing that goes beyond facts, rules, and the procedures of a rule-governed inquiry. Professional knowing is more than an ability to reflect on our actions. In professional action, knowing is in the action, involving a competence to deal with the often messy, indeterminate zones of practice with their uniqueness, uncertainty, and value conflicts. Learning a practice entails initiation into the traditions of a community of practitioners and the practice world they inhabit; this may take place in the direct exposition to real conditions and practices of work through apprenticeship, or through their simulation in a practicum, to be discussed below.

BOX 5.3   Interview Practica

I. Learn interviewing by witnessing others interviewing

Sit in with a more experienced researcher who is interviewing, listen and observe, and gradually become more active, taking care of the recording and participate as a co-interviewer.

LEARNING LESSONS:

— Significance of the social relationship of interviewer and interviewee
— Importance of interviewers being knowledgeable on the topics they are asking about
— Value of staging and scripting an interview in advance
— Differences between the live interview situation and the sound-recorded interview

---

II. Learn interviewing by practicing interviewing and listening to the interviews

LEARNING LESSONS:

— Acquiring self-confidence through gradual mastery of the practical, technical, social, and conceptual issues of interviewing, and thereby becoming able to create a safe and stimulating interview situation
— Options for improving the content, formulation, and sequence of the questions
— Becoming aware of the importance of mastering the art of second questions
— Videotaping some pilot interviews, which will heighten sensitivity to the body language of interviewer and interviewee

III. Learn interviewing in a community of interview researchers

LEARNING OCCASIONS:

— Listening to interviews and interviewer stories about interviews
— Obtaining an impression of the overall designing of an interview inquiry by observing several interviewers working at different stages of their projects
— Receiving feedback on your own pilot interviews from more experienced interviewers
— Having another researcher interview you about your research theme, and uncover your unreflected assumptions and personal biases about the interview theme

---

Inspired by Schön's practicum for educating the reflective practitioner, we have in Box 5.3 outlined interview practica, suggesting learning occasions and listing lessons to be learned. In the interview practica and in the transcription task above, the focus has been on learning two stages of an interview inquiry—interviewing and transcribing. In corresponding ways, the other stages of an interview inquiry, such as analysis, may be learned in an interview practicum or by apprenticeship in a research community, which provides a chance to observe and, to some extent also assist, more experienced performers of the trade.

Writing a book on the craft of interviewing involves a paradox of presenting explicit guidelines for a craft, which consists of practical skills and of personal know-how that often remains tacit and context bound. Research apprenticeships and practica in the interview craft, where students learn interviewing in the interaction of research communities with masters of the craft, are not commonly available. When the option is to be self-taught, a manual in a book form may be better than nothing, providing some explicit guidelines and rules of thumb for the beginner levels of interviewing skills. While one does not learn to become a good interviewer only through reading a book about interviews, a book can provide some information about the terrain through which an interview journey goes, and about available equipment and toolboxes for the trip, and thereby facilitate the journey and enhance the quality of the knowledge the interview traveler brings home.

Learning from examples of best practice by a careful reading of published interviews by masters of the trade—ranging from the clinical interviews in Rogers's *Client-Centered Therapy* (1956) to the research interviews in Bourdieu and colleagues' *The Weight of the World: Social Suffering in Contemporary Society* (1999)—can support acquiring the skills of the trade.

Rigorous training in the craft of research interviewing is not commonplace in social science methodology programs today. Perhaps as a consequence of an understanding of interviewing as a rule-governed method of data gathering, there is in academia little emphasis on extended personal training of interviewers. The road to mastery of the interview craft through a transcribing task, an interview practicum, or ideally a research apprenticeship, may appear too cumbersome and time consuming. Rather than such a slow learning process, fast learning in a weekend course and an introductory textbook may in some instances be considered sufficient to embark on a Ph.D. project based on interviews.

In contrast, intensive training of interviewers is required in several professional contexts. Thus therapeutic interviewers, as well as commercial interviewers, who live by the value of their interviews to their clients and employers, may need several years of training to master the craft of interviewing. Authorization to conduct psychoanalytic interviews thus requires years of practical and theoretical training. In the Hawthorne investigations of human relations in industry, several years of training were considered necessary to become a proficient interviewer. For focus group interviews in consumer research today, Chrzanowska (2002) has estimated that about two years of training is necessary for becoming a qualified focus group moderator. However, as Sennett (2006) has argued in his book on *The Culture of the New Capitalism,* it may be that the current consumer society is too "impatient" for people to be able to think of themselves as craftsmen. Sennett's point is not that people are too lazy to become capable craftsmen, but that "the economy creates a political climate in which citizens have difficulty in thinking like craftsmen" (p. 171). This, perhaps, is a specific danger in research, where researchers—novices and "masters" alike—are under pressure to deliver "results" quickly.

The professional examples of prolonged training should not lead anyone away from undertaking academic research interviewing. While such extensive training may not be realistic in a crammed university program, the practical training needed for acquiring advanced interviewing skills should not be overlooked. One may perhaps consider, however, within the limited time span in an academic program for learning research interviewing, reducing somewhat the time spent on paradigmatic discussions of the scientific status of interview research and paying more attention to learning the interpersonal practice of interviewing. A related suggestion emphasizing training the personal skills of research interviewers was put forth half a century ago by a research group whose work will be presented in the next chapter (in Box 6.6). They wrote:

> It is unfortunate that academic departments so often offer courses on the statistical manipulation of human material to students who have little understanding of the problems involved in securing the original data. Learning how to meet people of all ranks and levels, establishing rapport, sympathetically comprehending the significance of things as others view them, learning to accept their attitudes and activities without moral, social or esthetic evaluation . . . these are the elements to be mastered by one who would gather human statistics. When training in these things replaces or at least precedes some of the college courses on the mathematical treatment of data, we shall come nearer to having a science of human behavior. (Kinsey, Pomeroy, & Martin, 1948, p. 35)

PART II

SEVEN STAGES OF AN
INTERVIEW INVESTIGATION

———•◆•———

W e now turn from the conceptual issues of interview research to the practical steps of conducting an interview inquiry. We attempt to unravel the apparently mystical craft of interviewing into discrete steps, exemplify the steps, put forward some rules of thumb for the hands-on issues of the different steps, provide toolboxes on the way, and suggest learning exercises. We do not leave the epistemological and ethical issues raised in Part I behind, but continually refer to them in order to shed light on the concrete decisions to be made throughout an interview project. The chapters follow the temporal course of a qualitative interview investigation through seven stages: thematizing, designing, interviewing, transcribing, analyzing, verifying, and reporting. For each chapter, practical learning tasks are suggested in the appendix. The chapters present some of the multitude of courses an interview inquiry may follow, outline choices available, and discuss their consequences for the knowledge produced. We believe our depiction of an interview journey in terms of seven stages may be useful for novice researchers, but the capable craftsman may not always recognize these steps in his or her own practice, as practice is based more on well-formed habits and intuitive expertise than on explicit rules and guidelines.

SIX

THEMATIZING AND DESIGNING AN INTERVIEW STUDY

———————◆•◆•◆———————

‧

I n this chapter we outline a systematic planning of an interview study. We first present an account of the potential logistical and emotional hardships of an interview journey, showing how things can go wrong when the overall design of an interview project is not taken into account. Then, an idealized seven-stage route for an interview inquiry is outlined. Thereafter we treat more fully the preinterview stages of thematizing, or conceptualizing, and designing an interview project. We postulate that the better the preparation for an interview, the higher the quality of the knowledge produced in the interview interaction, and the more the postinterview treatment of the interviews will be facilitated. We conclude the chapter by addressing the issue of mixed methods in interview research.

SEVEN STAGES OF AN INTERVIEW INQUIRY

The open structure of research interviewing is an asset as well as a problem field in interview investigations. No standard procedures or rules exist for conducting a research interview or an entire interview investigation. There are, however, standard choices of approaches and techniques at the different stages of an interview investigation, and this chapter describes some of the choices to be made. The aim is to enable the interview researcher to make thoughtful decisions about method, based on knowledge of the topic of the study, the methodological options available, their ethical implications, and anticipated consequences of the choices for the entire interview project.

Articles in social science journals often give a rather formalistic picture of the research process as following a clear methodological procedure. Editorial requirements promote a distorted rationalized picture of scientific research as a logical, linear process—which is far from the continually changing actual research process with its surprises, design changes, and reformulations of concepts and hypotheses. A formalistic presentation of an interview investigation may also be required in applications for research funding, where the emphasis is on clear and well-structured methodological proposals. But such formal presentations of interview research are in most cases untrue to the often messy practice of conducting an interview study.

In this book, we aim to go beyond a dichotomy of all methods versus no method by focusing on the craftsmanship of the interview researcher. The openness of the interview, with its many on-the-spot decisions—for example, whether to follow up new leads in an interview situation or to stick to the interview guide—puts strong demands on advance preparation and interviewer competence. The absence of a prescribed set of rules for interviewing creates an open-ended field of opportunities for the interviewer's skills, knowledge, and intuition. Also, the design of an entire interview investigation follows fewer standard procedures than, for example, survey studies, where many of the method choices are already built into the standardized questionnaires and statistical analyses. While a qualitative interview is often characterized as "unstandardized," an entire interview investigation has often tended to be a rather standardized affair, going through five phases of hardships, as depicted in Box 6.1.

BOX 6.1   Emotional Dynamics of an Interview Journey

The Anti-Positivist Enthusiasm Phase. An interview project may start with enthusiasm and commitment. The researcher is strongly engaged in a problem and wants to carry out realistic natural life research. It is to be meaningful qualitative research of people's lives, and not positivist, quantified data gathering based on abstract theories.

The Interview-Quoting Phase. By now the researcher will have recorded the initial interviews and is intensively engaged in the interviewees' stories.

Forming a contrast to the ideological enthusiasm in the first phase, there is now personal involvement, a feeling of solidarity and identification with the subjects, who have revealed so much of their often oppressive life situation. At lunch the interviewer entertains his colleagues with a wealth of new quotations. Although exciting at first, it may be difficult after a while for the colleagues to remain fully involved in the myriad interview stories.

The Working Phase of Silence. After a time, silence falls upon the interview project. The researcher no longer brings up interview quotations at lunch. A colleague now asking about the project receives a laconic answer: "The interviews are being transcribed" or "The analysis has just started." This working phase is characterized by sobriety and patience.

The Aggressive Phase of Silence. A long time has passed since the interviews were completed and still no results have been presented. A colleague who now inquires about the project would run the risk of being met with distinct annoyance: The researcher bristles and more or less clearly signals, "It's none of your business." As for the researcher, this midproject crisis is characterized by exceeded time limits, chaos, and stress.

The Final Phase of Exhaustion. By now the interview project has become so overwhelming that there is hardly any time or energy left for reporting the originally interesting interview stories. In a common "save what can possibly be saved" termination, the interviews appear as isolated quotations in a report with little methodological and conceptual analyses. In cases where a more systematic "final report" does appear, the researcher may feel resigned because he or she has not succeeded in passing on to the readers in a methodologically justifiable way the original richness of the interview stories.

Box 6.1 depicts the emotional dynamics of an interview journey. The descriptions are based upon observations of colleagues and students undertaking interview projects, as well as recollections from the study of grading conducted by the first author. The emotional intensity of the hardship phases varies. Moments of enthusiasm, common at the beginning, can also occur in the later phases, such as when discovering new meanings through reading the interview transcripts. The description of the hardships of an interview project was first formulated some 25 years ago, at a time when interviewing was a relatively new method in the social sciences, with few conventions and little literature on interview research. The hardship tale may perhaps be less valid for interview research today.

BOX 6.2   Seven Stages of an Interview Inquiry

1. Thematizing. Formulate the purpose of an investigation and the conception of the theme to be investigated before the interviews start. The *why* and *what* of the investigation should be clarified before the question of *how—* method—is posed.

2. Designing. Plan the design of the study, taking into consideration all seven stages of the investigation, before interviewing (Chapter 6). Designing the study is undertaken with regard to obtaining the intended *knowledge* (Chapter 3) and taking into account the *moral* implications of the study (Chapter 4).

3. Interviewing. Conduct the interviews based on an interview guide and with a reflective approach to the knowledge sought and the interpersonal relation of the interview situation (Chapters 7, 8, and 9).

4. Transcribing. Prepare the interview material for analysis, which generally includes a transcription from oral speech to written text (Chapter 10).

5. Analyzing. Decide, on the basis of the purpose and topic of the investigation and of the nature of the interview material, which modes of analysis are appropriate for the interviews (Chapters 11, 12, 13, and 14).

6. Verifying. Ascertain the validity, reliability, and generalizability of the interview findings. Reliability refers to how consistent the results are, and validity means whether an interview study investigates what is intended to be investigated (Chapter 15).

7. Reporting. Communicate the findings of the study and the methods applied in a form that lives up to scientific criteria, takes the ethical aspects of the investigation into consideration, and results in a readable product (Chapter 16).

As a step toward preventing the events in the hardship tale from occurring, the interview investigation outlined in Box 6.2 provides a linear progression through seven stages from the original ideas to the final report. This idealized sequence may assist the novice interview researcher through the potential hardships of a chaotic interview journey and contribute to the researcher's retaining the initial vision and engagement throughout the investigation.

Some interview researchers advocate a more flexible approach to design than starting with a definite plan to be followed throughout the project. Rubin and Rubin (2005) thus suggest an iterative design, in which the researcher is continually adapting to new circumstances in the field, changing selection of subjects and questions on the way. In this form of responsive interview design, the research questions are generally open. If hypotheses are stated at the beginning, they may be modified or dropped as the project proceeds. As a contrast to the present somewhat formalistic and prescriptive depiction of an interview inquiry in seven stages, in Box 6.3 we present Bourdieu's more craftsmanlike conception of research interviewing, which was excerpted in the interview in Box 1.3.

**BOX 6.3   Bourdieu With Respect to Seven Stages in an
Interview Project**

1. Thematizing. "Researchers have some chance of being truly equal of their task only if they possess an extensive knowledge of the subject, sometimes acquired over a whole lifetime of research, and, also, more directly, through earlier interviews with the same respondent or with informants" (Bourdieu et al., 1999, p. 613).

2. Designing. Ruled by the principle of nonviolent communication: "For social proximity and familiarity provide two of the conditions of 'nonviolent' communication . . . So we left investigators free to chose their respondents from among or around people they knew or people to whom they could be introduced by people they knew (p. 610).

3. Interviewing. "It is not simply a question [of] collecting 'natural discourse' as little affected as possible by cultural asymmetry; it is also essential to construct this discourse scientifically, in such a way that it yields the elements necessary for its own explanation" (p. 611).

4. Transcribing. "Transcription then, means writing, in the sense of rewriting. Like the transitions from written to oral that occurs in the theater, the transition from the oral to the written, with the changes in medium, imposes infidelities which are without doubt the condition of true fidelity" (p. 622).

(Continued)

(Continued)

5. Analyzing. "Understood in this way, conversational analysis [the term is used here in a more general sense than we use it in discussing the methodic form of conversation analysis in Chapter 13 of this book] reads in each discourse not solely the contingent structure of the interaction as a transaction, but also the invisible structures that organize it . . . [it] avoids reducing . . . as in so many 'tape recorder sociologies,' and knows how to read in their words the structure of the objective relations, present and past, between their trajectory and the structure of the educational establishments they attended, and through this, the whole structure and history of the teaching system expressed there" (p. 618).

6. Verifying. "Rigor . . . lies in the permanent control of the point of view, which is continually affirmed in the details of the writing" (p. 625).

7. Reporting. "There is doubtless no writing more perilous than the commentary, which a public writer must provide for the messages that have been confided to them" (p. 625).

SOURCE: Excerpts from Pierre Bourdieu et al. Translated by Priscilla Parkhurst Perguson and Others. *The Weight of the World* © 1999 Polity Press for translation; 1993 Editions du Seuil.

In Box 6.3, we have organized Bourdieu's post hoc description of the conception of interviewing in *The Weight of the World* (pp. 610–625) into the format of the seven stages of an interview inquiry (the comparison of Bourdieu's and Kvale's approaches to interviewing is inspired by Glasdam, 2007). While corresponding in important aspects to the present depiction of interviewing, Bourdieu's approach is tied to the specific project and rests on a more dynamic and flexible approach to interview research. Whereas experienced interview researchers will tend to adapt more open and flexible designs, we will in this book for didactic purposes give a more idealized and formal presentation, which we believe may be useful to novice interview researchers. We now turn to the two preinterview thematizing and designing phases of an interview inquiry (for designing qualitative projects in general, we refer the reader to Flick, 2007, and Marshall and Rossman, 2006).

THEMATIZING AN INTERVIEW STUDY

A significant part of an interview project should take place before the sound recorder is turned on for the first actual interview. *Thematizing* refers to the formulation of research questions and a theoretical clarification of the theme investigated. The key questions when planning an interview investigation concern the why, what, and how of the interview:

- *why:* clarifying the purpose of the study
- *what:* obtaining preknowledge of the subject matter to be investigated
- *how:* becoming familiar with different techniques of interviewing and analyzing, and deciding which to apply in order to obtain the intended knowledge

The term *method* originally meant the way to the goal. In order to find or to show someone the way to a goal, one needs to know what the goal is. It is necessary to identify the topic and the purpose of an interview investigation in order to make reflected decisions on which methods to use at the different stages on the way to the goal. Consultations on interview projects may sometimes take the form of an explorative "counter"-interview. The counselor first needs to explore, by carefully questioning the novice interviewer, where the interview journey is heading, what is the research topic of the interview study, and why it is undertaken, before the many technical questions about methods can be addressed. There is a standard reply to the method questions about design of qualitative interview studies—the answer depends on the purpose and the topic of the investigation. The thematic questions of "why" and "what" need to be answered before the technical "how" questions of design can be posed and addressed meaningfully.

Research Purpose. Thematizing an interview study involves clarifying the purpose of the study—the "why." Interviews may be conducted to obtain empirical knowledge of subjects' typical experiences of a topic, such as in the phenomenological interview about learning reported in Chapter 2. Or interviews may seek knowledge of a social situation, such as in the three interview passages in Chapter 1, investigating the influence, respectively, of grading and of educational reforms on the learning and work situation in school (Boxes 1.1 and 1.2) and of the life situation of the downtrodden in France (Box 1.3).

Or an interview inquiry may study life histories, such as in biographical interviews, or historical events. In the case of oral history interviews, it is less the subject's experiences as such, which are of interest, than the information they provide about social and historical events. The interviews may also go further than charting subjects' experiences, or using the subjects as informants about events, and attempt to get beyond the self-presentations of the subjects and critically examine the personal assumptions and general ideologies expressed in their statements, such as the epistemic interviewing depicted in Chapter 2.

Interviews can have explorative or hypothesis-testing purposes. An *exploratory* interview is usually open, with little preplanned structure. In this case the interviewer introduces an issue, an area to be charted, or a problem complex to be uncovered, then follows up on the subject's answers and seeks new information about and new angles on the topic. Interviews that *test hypotheses* of group differences tend to be more structured; the wording and sequence of questions may here be more standardized in order to compare the interviews from the groups. Hypotheses testing can also take place within a single interview, where the interviewer questions in order to test hypotheses about a subject's conception of an issue, such as in the confrontational interview from Bellah and colleagues (1985) about commitment and individualism (Chapter 3).

Interviews can also be primarily descriptive and seek to chart key aspects of the subject's lived world. They may further attempt to develop theoretical conceptions of a topic, such as in the grounded theory approach developed by Glaser and Strauss (1967), to *inductively* develop an empirically grounded theory through observations and interviews. An interview investigation may also seek to *deductively* test the implications of a theory. And interviews can be conducted in order to develop knowledge for, and through, collective activities in action research. Interviews are also used as *background material* for further practical and theoretical studies. Schön's (1987) analysis of the reflecting practitioner is thus based on interviews with professionals, and Sennett's (2004) book on respect builds on extensive interview experience.

Subject Matter Knowledge. Thematizing an interview study involves clarifying the theme of the study—the "what." This entails developing a conceptual and theoretical understanding of the phenomena to be investigated in order to establish the base to which new knowledge will be added and integrated. Familiarity with the theme investigated is required to be able to pose relevant questions, whether they concern the essence of beauty, truth, and goodness in a Socratic dialogue, the strategy of a master chess player, or trends in rap music.

The thematic focus of a project influences what aspects of a subject matter the questions center upon, and which aspects remain in the background. The influence of differing theoretical conceptions upon choices of method may be exemplified by an imagined interview with a pupil about the *meaning of teasing*. Different psychological theories lead to different emphases on emotions, experiences, and behavior—as well as on the temporal dimensions of past, present, and future. Say a school psychologist is interviewing a pupil who, according to a teacher's complaints, is continually teasing the other pupils and thereby disturbing the class. The interview might be conducted from a Rogerian client-centered approach, a Freudian psychoanalytic approach, and a Skinnerian behavior-modification approach, respectively. Different kinds of interview questions are required to obtain the kinds of information necessary to interpret the meaning of teasing with respect to these theories. They would, simplified here, focus on present experiences and feelings about teasing, on family history and emotional dynamics, and on future behavioral consequences, such as the reactions of fellow pupils to the teasing, respectively. If these theoretical approaches, which highlight different aspects of the meaning of teasing, are not introduced until the analysis stage, the interviews may lack the relevant information for making specific interpretations on the basis of the different theories.

One definition of science is the systematic production of new knowledge. Without any presentation of the existing knowledge about the topic of an investigation, it is difficult for both researcher and reader to ascertain whether the knowledge obtained by the interviews is new, and thus what the scientific contribution of the research is. A theoretical naiveté of many current interview projects is not intrinsic to qualitative research. The contributions of Freud, as well as later psychotherapists, and of Piaget, testify to the potentials of theorizing on the basis of qualitative interviews. More recent examples include the studies of Hargreaves and Bourdieu (Boxes 1.2 and 1.3) in which the interviews were based upon, and served to develop, theoretical understanding. Bourdieu mentions a problem when his research group attempted to use some of the young people studied to interview each other. Whereas the young people would connect well with each other and thus interview in line with the principle of nonviolent communication of the study, the resulting interviews demonstrated that they lacked the theoretical knowledge necessary for conducting penetrating interviews.

Familiarity with the content of an investigation is not obtained only through literature and theoretical studies. Just hanging out in the environment where the interviews are to be conducted will provide the interviewer with an introduction to the local language, the daily routines, and the power structures, and so provide a sense of what the interviewees will be talking about, a situation not unlike that of the novice apprentices in a workshop. Familiarity with the local situation may also sensitize the researcher to local ethicopolitical issues of the community, which need be taken into account when interviewing and reporting the interviews. Particularly for anthropological studies, a familiarity with the foreign culture is required for posing questions, as testified by the anthropologist Jean Lave:

> One of the reasons for doing field trips is that you are presented with how abstract is the most concrete of your concepts and questions when you are at home in the library. When I first went to Brazil I made my way 2,000 miles into north central Brazil and I arrived in a small town. I heard that there were Indians who actually were in town. And I can remember an incredible sense of excitement. I rushed out and walked around town until I found this group of Indians and walked straight up to them—and then I didn't know what to say. I wanted to ask: "Have you got moiety systems?" (a special kind of kinship relations). And it didn't make sense to do that. In fact it took four months to find a way to ask a question with which I could discover from people whether they did have moiety systems. (Lave & Kvale 1995, p. 221)

Thematizing in the Grade Study. We will now exemplify thematizing in an interview investigation of the study of grading in Danish high schools from 1978 (an interview sequence from the study was presented in Box 1.1, and the overall design follows in box 6.4). The study was instigated by a public debate about the effects of grading in connection with a new policy of restricted admission to college based on grade point averages from high school (Kvale, 1980). I (Steinar Kvale) had been involved in a newspaper debate with the Danish Minister of Education, who maintained that there would be hardly any educational or social impact from a restricted university admission based on grade point averages, and that contrary findings from other countries could not be generalized to the Danish situation. (The following sections are in the first person singular form, since this material is based on my personal experiences.) I decided to investigate the issue empirically by asking Danish pupils and teachers about their experiences with grading. Several hypotheses were formulated in advance. The first hypothesis of a grade perspective was that grading influences the process of learning and the social situation where learning occurs. A second hypothesis stated that the prevalence of the grading perspective would increase with a restricted admission to college based on grade point averages. The hypotheses were based on research literature, as discussed in my

Ph.D. dissertation on examinations and power (Kvale, 1972) and on later investigations of grading by other researchers.

DESIGNING AN INTERVIEW STUDY

Designing an interview study involves planning the procedures and techniques—the "how"—of the study. The following comments on formal designing pertain more to larger systematic interview investigations than to smaller flexible exploratory studies. And an experienced interview researcher may obtain important knowledge using more informal interview approaches.

Designing the Grade Study. As an example of designing a large interview study, we will outline in seven stages the first author's interview study of grading in Danish high schools.

BOX 6.4   Seven Stages of a Grade Study

1. Thematizing. Formulation of hypotheses about the influence of grading on pupils on the basis of previous studies.

2. Designing. Planning the interviews with 30 high school pupils and 6 teachers.

3. Interviewing. A detailed guide was used for the individual interviews, each of which lasted about 45 minutes and was tape-recorded.

4. Transcribing. All 36 pupil and teacher interviews were transcribed verbatim, resulting in about 1,000 pages of transcripts.

5. Analyzing. The 30 pupil interviews were categorized with respect to different forms of grading behavior. The interviews with the pupils and the teachers were also subjected to more extensive qualitative interpretations.

6. Verifying. Reliability and validity checks were attempted throughout the project, including interviewer and scorer reliability, and validity of interpretations.

7. Reporting. The results were reported in a book and in journal articles.

SOURCE: *Spillet om karakterer i gymnasiet: Elevinterviews om bivirkninger af adgangsbegrænsning* (The Grading Game in High School: Pupil Interviews About Side Effects of Restricted Admission), by S. Kvale, 1980, Copenhagen: Munksgaard.

Box 6.4 depicts post hoc, in an orderly fashion, the seven stages of the, at times chaotic, investigation of grades. I took special care to have a methodologically well-controlled design for the interviews about grades because their influence was a controversial public topic at the time, and I had myself a rather critical view of using the grade point average for admissions to college. The use of qualitative interviews in research was also fairly new and contested at the time. Thirty high school pupils were interviewed individually about their experiences with grades. This number was a compromise between obtaining a representative sample and the resources available for the study. In order to counteract possible special circumstances at a single high school, the pupils were selected at random from three schools, one class in each. To counteract individual interviewer bias, the 30 pupils were distributed among four interviewers, three student assistants and myself. In order to gain an alternative perspective on the effects of grading, six teachers were also interviewed.

The Temporal Dimension of an Interview Design. The temporal dimension of an interview investigation should be kept in mind from the first thematizing to the final reporting stage, taking into account the interdependence of the seven stages. This includes having an overview of all seven stages of an interview study, paying attention to the interdependence of the stages, and also pushing forward tasks at later stages—such as analysis and verification—to earlier stages, returning spirally to earlier stages, keeping the end point of the study in sight and taking into account that the interviewer may become wiser throughout the study.

- *Overview.* A key factor in an interview inquiry is to develop an overview of the entire investigation before starting to interview. When using the more standardized methods, such as experiments, questionnaires, and tests, the very structure of the instruments requires advance decisions about the way in which the study will be conducted. In this case, methodological alternatives are already built into the instruments, for instance by the response alternatives of questionnaires and by computer programs for statistical analysis and presentation of the numerical findings. In open and nonstandardized interview studies, however, the choices of method may first appear during the investigation, in some cases when it is too late to make decisions appropriate for the topic and purpose of the study.

- *Interdependence.* There are strong interconnections among the choices of method made at the different stages. A decision at one stage has consequences that both open and limit the alternatives available at the next stage. For example, a statistical generalization of the findings of an interview study to larger groups will require that certain criteria—regarding size and representativity of the sample of subjects—are taken into account already by selection of interviewees. And if the researcher wants to make a systematic linguistic or conversational analysis of the interviews, it will not be possible or will require a time-consuming retranscription if the interviews had been edited into normal English by the transformation from oral to written language.

- *Push Forward.* An interviewer may attempt to do much of the work of the postinterview stages at earlier stages. Although the problems of an interview project tend to surface in the analysis stage, more often than not they originate in earlier stages. The solution is to improve the quality of the original interviews. Thus, clarifying the meanings of statements during an interview will make the later analysis easier and more well founded; asking control questions during the interview will facilitate the validation of interpretations. Improving interview quality is not simply a question of better interview techniques; it also involves a reflective thematization of the topic and purpose of the inquiry from the very beginning.

- *Spiraling Backwards.* An interview project is often characterized by a back and forth process between different stages. The linear progression of the seven stages discussed here may in practice be modified into a circular or spiral model where the researcher, with an extended understanding of the themes investigated, at later stages returns to earlier stages. This may be to reinterview or interview new subjects, to retranscribe specific passages, or to reanalyze the interview stories from new perspectives.

- *Keep the Endpoint in Sight.* From the start of the investigation, keep the expected result in view. Is the intended outcome a Ph.D. thesis or an internal evaluation report? Will a publication result from the study? As a short article or as a book? For a scientific forum or for the general public? The answers to such questions should serve as guidelines throughout the stages of the interview project, assisting the informed decisions made on the way and keeping the project on track toward the goal. The nature of the final report is decisive for decisions at earlier stages on such ethical issues as informing the interviewees about later use of what they say, obtaining written permission to quote extensively from their interviews, and judging how personal themes can be addressed in interviews intended for public use.

• *Getting Wiser.* An interviewer may learn throughout an investigation. The conversations with the subjects can extend and alter his or her understanding of the phenomena investigated. The interviewees bring forth new and unexpected aspects of the phenomena studied; and during analysis of the transcribed interviews new distinctions may be discovered. This is in line with the purpose of an exploratory study—to discover new dimensions of the research topic. In comparative hypothesis-testing studies, however, it may create design problems if the interviewer continually obtains new significant insights. Novel dimensions of a phenomenon may be discovered in the middle of a series of interview testing, for example, the influence of grades upon girls' and boys' ways of learning a school subject. The dilemma will then be whether to improve the interview guide with specific questions to the new dimensions in the remainder of the interviews, and not have comparable groups, or to refrain from using the new insights in the remaining interviews. There is no easy solution to the dilemma of becoming wiser as a threat to standardized conditions, except to be as clear as possible about the main purposes of a study from its inception. Thus in exploratory studies the questioning may continually improve as the researcher learns more about a topic, ideally resulting in a sophisticated form of interviewing that is receptive to the nuances and complexities of the topic explored.

• *Time and Resources.* At the start of a project the resources and the expertise necessary for an interview study can be easily overlooked. Relevant questions about resources are: How much time does the researcher have available for the study? Is there any money available for assistance—for example, for transcribing the interviews? Usually, conducting the interviews themselves is not time consuming, whereas transcribing them requires much more time, and assistance for transcription may be expensive. The subsequent analysis of the transcripts is more often than not the most time-consuming part of the interview study.

• *Work-Journal.* In order to keep track of the temporal vicissitudes of an interview journey the researcher can keep a work-journal as a record of his or her learning throughout the investigation. The daily insights obtained by the construction of interview knowledge are noted down, including altered understandings of previous experiences, as well as reflections on the research process. During analysis, verification, and reporting, the work-journal will then provide the researcher with a frame for understanding and reflecting on the processes and changes in the knowledge production throughout an interview inquiry.

How Many Interview Subjects Do I Need? The answer to this common question is simply: "Interview as many subjects as necessary to find out what you need to know." In qualitative interview studies, the number of subjects tends to be either too small or too large. If the number of subjects is too small, it is difficult to generalize and not possible to test hypotheses of differences among groups or to make statistical generalizations. If the number of subjects is too large, there will hardly be time to make penetrating analyses of the interviews.

The number of subjects necessary depends on the purpose of a study. If the aim is to understand the world as experienced by one specific person, say in a biographical interview, this one subject is sufficient. If the intention is to explore and describe in detail the attitudes of boys and girls toward grades, new interviews might be conducted until a point of saturation, where further interviews yield little new knowledge. If the goal is to predict the outcome of a national election, a representative sample of about 1,000 subjects is normally required, so qualitative interviews are out of the question, and questionnaires are used. If the purpose is to statistically test hypotheses about different attitudes of boys and girls toward competition for grades, the necessary sample may be as small as three boys and three girls. Depending on the distribution of the findings, a Fisher test of significant differences between the two groups of three can be made at a probability level of $p < .05$ (Siegel, 1956).

In common interview studies, the number of interviews tends to be around 15 +/– 10. This number may be due to a combination of the time and resources available for the investigation and a law of diminishing returns (beyond a certain point, adding more respondents will yield less and less new knowledge). A general impression from current interview studies is that many would have profited from having had fewer interviews in the study, and instead having taken more time to prepare the interviews and to analyze them. Perhaps as a defensive overreaction, some qualitative interview studies appear to be designed on a misunderstood quantitative presupposition—the more interviews, the more scientific.

We will again draw in the grade study to illustrate the issue of the number of subjects and also the use of mixed methods optimizing the strong points of interviews and questionnaires. In the interview passage, a pupil asserted a connection between pupils' talkativity and their grades (Box 1.1). In a follow-up study by two of the students involved in the research project, the statement was split into two items in a questionnaire (Hvolbøl & Kristensen, 1983).

Table 6.1   From Interview Statements to Questionnaire Items

Interview Statement				
Pupil: Grades are often unjust, because very often—very often—they are only a measure of how much you talk, and how much you agree with the teacher's opinion.				
	Percentage of 239 Pupils			
Questionnaire Items	Strongly Agree	Agree	Disagree	Strongly Disagree
Grades are often an expression of how much one talks in class	20	62	15	3
Grades are often an expression of how much one follows the teacher's opinion	4	20	57	19

Table 6.1 presents the percentage of agreement among 239 pupils from six schools when the interview statement from Box 1.1 was converted into a questionnaire form. It turned out that a clear majority of the pupils agreed with the first part of the statement of the pupil interviewed—that grades are an expression of how much one talks, whereas a majority disagreed with the second part—that grades are often an expression of how much one goes along with the teacher's opinion. With the large number of subjects, the questionnaire could then check the generality of the views stated by one pupil, a generality which would require too many resources to test by the more time-demanding qualitative interviews. In retrospect, the grade study would probably have produced more valuable knowledge with fewer, but longer, more intensive interviews and by subjecting the interviews to more penetrating interpretations. The questionnaire developed later on the basis of the interviews sufficed to test the generality of the interview findings.

This example points to strengths and weaknesses of interviews and questionnaires. The interview brought out interesting beliefs about which behaviors lead to good grades, whereas the questionnaire made it possible to test how prevalent these beliefs were among a large number of pupils. While the interviewer could closely question the strength of a pupil's belief, and might also have obtained concrete examples supporting the claims, the questionnaire used did not follow up the pupils' statements. In the present approach there are no important epistemological differences between mixing the methods of interviews and questionnaires; they merely provide answers to different kinds of research questions, such as what kinds of beliefs the pupils have about grades and how many pupils have these kinds of beliefs.

When Not to Interview. In this chapter on interview design it may be pertinent also to mention some areas and uses for which qualitative interviews are little suited. Before embarking on a research project using qualitative interviewing, one should ask whether this will be the most adequate way of answering the research questions that one is interested in. One purpose of the present book is to lead some readers *away* from using research interviews, by pointing out that in several cases other methods may be more appropriate for the subject matter and purpose of the intended research.

If a study seeks to predict the behavior of larger groups, such as voting behavior, larger samples of respondents are necessary than would be possible to cover with time-consuming qualitative interviews; in such cases survey questionnaires with precoded answers are the most relevant method. Also, when there is little time available for a project, questionnaires are a better choice because they are usually faster to administer, analyze, and report than qualitative interviews.

If you want to study people's behavior and their interaction with their environment, the observations and informal conversations of field studies will usually give more valid knowledge than merely asking subjects about their behavior. If the research topic concerns more implicit meanings and tacit understandings, like the taken-for-granted assumptions of a group or a culture, then participant observation and field studies of actual behavior supplemented by informal interviews may give more relevant information than formal interviews.

If the purpose of a study is to obtain penetrating personal knowledge of a subject, then this may best be obtained through the trust developed in the close, personal interaction developed through a long and emotional therapy process (see the non-directive therapeutic interview and the depiction of the psychoanalytic interview in Boxes 2.5 and 2.6). Whereas the challenges to a person's established self-image and the provocation of strong feelings are necessary parts of therapy, inciting such intensive emotional reactions only for research interests would be unethical.

Potter and Hepburn (2005) have criticized the widespread use of interviews in qualitative psychological research, and they question the often taken-for-granted choice to do interviews (p. 283). From a discourse theoretical point of view, they favor analyses of naturally occurring talk and action over the practice of qualitative research interviewing. They argue that doing interviews involves several risks that are not always thematized, such as flooding the interaction with social science agendas and categories, a bias toward cognitive and individualist explanations, and an undertheorized relationship of the varying positions between interviewer and interviewee. We agree that the choice to do interviews should always be based on good reasons that are ideally explicated in the interview report, and also that greater attention should be directed at the practice of interviewing itself, including the positions of researcher and respondent, with regard to the implications for the knowledge produced.

When planning an interview study, it may thus be appropriate to consider whether other methods may be more suitable for the theme and purpose of the project. This being said, it should not be forgotten that interviews are particularly well suited for studying people's understanding of the meanings in their lived world, describing their experiences and self-understanding, and clarifying and elaborating their own perspective on their lived world.

MIXED METHODS

Today, the use of "mixed methods" has become a controversial theme, in particular the combination of qualitative and quantitative methods with allegedly different paradigmatic assumptions. The controversy is embedded in the sociopolitical context of the social sciences, where a methodological hierarchy is often implied, with quantitative methods at the top and qualitative methods relegated to a largely auxiliary role (Howe, 2004).

Somehow the use of mixed methods, and their allegedly paradigmatic differences, was not an issue in the historical interview studies mentioned earlier. Thus Piaget's investigations of children's thought were a free combination of observations, quasi-naturalistic experiments, and interviews. The many thousand interviews in the Hawthorne study were conducted to find out why production continued to increase with experimental changes in the workers' conditions and how the human relations could be improved in the factories. In current market research, a combination of focus group interviews and questionnaires is rather a matter of course when introducing a new product, and in the marketing of politicians focus groups and opinion polls are standard techniques.

Interviews are often applied in *case studies,* which focus on a specific person, situation, or institution. Interviews can also serve as an *auxiliary method* in conjunction with other methods. In participant observation and in ethnographic field studies, more or less informal interviews are thus important sources of information. In the construction of questionnaires it is common to use pilot interviews to chart the main aspects of a topic and also test how survey questions are understood. In postexperimental interviews, subjects are questioned on how they understood the experimental design.

In the present pragmatic approach, the different methods are different tools for answering different questions; qualitative methods refer to *what kind,* and quantitative methods to *how much of a kind.* The important issues of mixed methods are less on a paradigmatic level than on a practical level. Thus, to work with different media—words and numbers—and to manage questionnaire construction and statistical analyses, as well as interview forms and analytic techniques, all at a high level of quality, demands expertise obtained through long training. We described in Chapter 5 how questionnaires and qualitative interviews use different logics, with the administration of questionnaires following mechanically standard rules with a minimum of personal judgment, and qualitative interviews resting upon the interviewers' skills and situated personal judgment in the posing of questions.

We conclude this section on mixed methods by presenting two boxes, with a constructed example and a historical example of the use of mixed methods, respectively. The first concerns a forced melange of different methods, the second a deliberate use of different methods.

BOX 6.5   Qualitative Research Proposals in the Crossfire

Qualitative interviews and field studies involve specific research approaches. Their techniques and logics differ from those of survey studies, tests, and experiments. In some university departments, the committees that accept or reject Ph.D. proposals may, however, regard the logic of the experiment,

(Continued)

(Continued)

often as codified in a methodological positivist tradition, as the basic scientific method to which qualitative projects have to adapt. Transferal of qualitative projects to a mainstream methodological logic may be a self-defeating project, as hopeless as playing according to the rules of European soccer in a game of American football.

Qualitative research proposals are simply met by some Ph.D. review boards with requirements that are alien to good qualitative research. For example, an exploratory qualitative project, valuing openness to an area, may be rejected on the grounds that there are not sufficiently precise hypotheses for the project, or that a control group is lacking. A qualitative project may be allowed to work qualitatively all the way in some disciplines, such as in most anthropological studies, whereas review boards in other social sciences may require that qualitative interviews are followed by quantitative categorizations or with questionnaires to larger groups for quantification and generalization of the findings. A qualitative project in a sensitive area may need complex and flexible modes of recruiting willing subjects, whereas a mainstream committee may require a random selection of subjects. The enforcement of such methodological demands, alien to qualitative research, may simply lead to low quality qualitative research.

In Box 6.5 we have constructed a case of an enforced use of mixed methods. While such contradictory demands may appear absurd in anthropology, they occur in some departments of psychology. Qualitative projects that are forced to play two different ball games at the same time, and adapt to the logic of survey studies, are likely to result in bastards of low quality, according to quantitative as well as qualitative logics.

Although it is a made-up case, it exemplifies a general bureaucratization of research in line with methodological positivism that comes close to method requirements, which today also follow from evidence-based practice (see Box 3.1).

We do not have any easy solution for the dilemma of students interested in qualitative research who are doing research in methodologically conservative university departments. Mainstream review boards are parts of powerful social practices, where divergent epistemological arguments may have little say. Transferal to departments or universities open to qualitative research may not always be a realistic way out. Here we give some suggestions for qualitative proposals in methodologically modestly open departments:

- Make the research proposal clear and transparent, conceptually precise, and realistic.

- Demonstrate a thorough knowledge of the disciplinary literature on your research topic.

- Demonstrate knowledge of the methodological literature of qualitative, as well as of mainstream, lines of a discipline. Argue briefly and concisely why, for example, a precise hypothesis or random sampling would be at odds with the purpose and logic of the design, preferably with references to both lines of method literature in the discipline.

- Demonstrate, when possible, that similar designs have already been used in well-recognized studies in your field.

- When possible, refer to your own previous experience with the qualitative methods you want to apply, and ideally to your publications in the field.

We conclude this section on mixed methods by depicting a historical interview study, which in innovative ways followed the respective logics of qualitative interviewing and statistical data analysis, and produced significant and provocative new knowledge.

BOX 6.6   The Kinsey Interview Study of Sexual Behavior

The study *Sexual Behavior in the Human Male* (Kinsey et al., 1948) was designed to accumulate a body of scientific facts that could provide the basis for scientific generalizations about sexual behavior. To this purpose, the research group interviewed about 6,000 men for an hour or a more about their sexual behaviors. The statistical results were rather schocking at the time, as they treated subjects that had previously been taboo, and they challenged conventional beliefs about sexuality by documenting a pervasive

(Continued)

(Continued)

occurrence of sexual behaviors, many of which were considered immoral and illegal at the time, such a premarital intercourse, masturbation, oral-genital contacts, and homosexuality.

The book contains a unique discussion of the human interaction in the qualitative research interviews, which constituted the original data for the scientific generalizations, and also a sophisticated coding procedure for converting the subjects' answers into quantitative data for statistical analysis. We recommend the book for reading about how to address intimate and taboo aspects of private lives in a humane, ethical, and methodologically sophisticated way.

Kinsey and his colleagues introduce their chapter on interviewing by stating that the quality of a case study begins with the quality of the interviewing by which the data have been obtained. When establishing rapport with the interviewee, "The subject should be treated as a friend or a guest in one's home" (p. 48). Rapport was created by being fully present to the subjects, being honest with them, and showing sincere appreciation for their help in securing stories. The researchers discuss in detail how to put their subjects at ease, assuring privacy, the sequencing of sensitive topics, and so on. The sympathetic interviewer convinces the subject that he is anxious to comprehend what his experience has meant to him. If the interviewer succeeds, the subject is willing to tell his story: "The interview has become an opportunity for him to develop his own thinking, to express to himself his disappointments and hopes, to bring into the open things that he has previously been afraid to admit to himself, to work out solutions to his difficulties. He quickly comes to realize that a full and complete confession will serve his own interests" (p. 42). The authors emphasize that it is difficult to explore phenomena efficiently unless one has some understanding of the sort of thing that one is likely to find. In this case, the interviewer needed to comprehend the whole range of techniques in each possible type of sexual behavior. The interviewer needed to understand and be able to converse naturally about socially taboo and illegal sexual activities. In addition there was the need to learn by heart the complex coding scheme used with about 500 items. It was estimated that a full year of training was required before the interviewer was ready to conduct an interview.

The researchers took great care to protect the confidentiality of their subjects, since many of the sexual behaviors interviewed about, could, if disclosed, lead to long-term prison for the subjects in the middle of the

20th century in the midwestern United States. The interviews were not recorded by tape or in longhand, but were coded during the interview, and a cryptographer was used to prevent any identification of the subjects. We will return to the coding during the interview in Chapter 12.

Box 6.6 depicts a historical example of a purposeful use of mixed methods, with the development of sensitive qualitative interviewing in a taboo area, as well as a meticulous coding and statistical analysis of the interview findings. Here is no question of mixed logics, the qualitative logic of interviewing and statistical reasoning function each in their own right. The controversial findings of the study have had a marked effect on the public understanding of sexuality and morality.

ᴴ SEVEN ᴷ

CONDUCTING AN INTERVIEW

———•◆•———

W e now turn to the actual process of producing knowledge through an interview. This chapter exemplifies interviewing by means of a demonstration interview, and discusses the interaction and the questioning in the interview. We take up preparations for an interview, such as setting the stage for the interview and preparing a script in the form of an interview guide. We also deal in some detail with researcher questions, interviewer questions, the linguistic forms of questions, and the art of second questions. We do not propose any general rules for interviewing, but describe some of the techniques of the interview craft. With the techniques of interviewing mastered, the interviewer may concentrate on the subject and the subject matter of the interview: "Only . . . *a reflex reflexivity* based on a craft, on a sociological 'feel' or 'eye,' allows one to perceive and monitor *on the spot,* as the interview is actually taking place, the effects of the social structure within which it is occurring" (Bourdieu et al., 1999, p. 608).

The research interview is an interpersonal situation, a conversation between two partners about a theme of mutual interest. In the interview, knowledge is created "inter" the points of view of the interviewer and the interviewee. The conversations with the subjects are usually the most engaging stage of an interview inquiry. The personal contact and the continually new insights into the subjects' lived world make interviewing an exciting and enriching experience. In this chapter, we show in some detail the application of a semi-structured life world interview, which we already addressed in

Chapter 2 in relation to phenomenological philosophy. In the following chapter, we present an overview of a variety of other interview forms by means of short examples. The semi-structured life world interview seeks to obtain descriptions of the life world of the interviewee with respect to interpreting the meaning of the described phenomena; it will have a sequence of themes to be covered, as well as some suggested questions. Yet at the same time there is openness to changes of sequence and forms of questions in order to follow up the specific answers given and the stories told by the subjects. The open phenomenological approach of a life world interview to learning from the interviewee is well expressed in this introduction to anthropological interviewing:

> I want to understand the world from your point of view. I want to know what you know in the way you know it. I want to understand the meaning of your experience, to walk in your shoes, to feel things as you feel them, to explain things as you explain them. Will you become my teacher and help me understand? (Spradley, 1979, p. 34)

A CLASS INTERVIEW ABOUT GRADES

Below is an interview that the first author conducted before a class at an interview workshop at Saybrook Institute, San Francisco, in 1987. Although the interview situation is artificial, it gives in a condensed form a fair picture of a semi-structured life world interview. The interview is reproduced shortened and verbatim, with only a few minor changes in linguistic style.

Steinar Kvale 0: I will now attempt to demonstrate the mode of understanding in a qualitative research interview, and I need a volunteer. It will be a rather neutral topic, it's not a psychoanalytic depth interview. The interview will take about 10 minutes and afterwards we will discuss it here.

A woman in her 30s volunteers.

SK 1: Thank you for your willingness to participate and be interviewed here. I have been studying the effects of grades in Europe for some years, and now I'm interested in the meaning of grades for American students and pupils. I want to first ask you a maybe difficult question. If you'll try to remember back when you went to primary school, are you able to remember the first time you ever had any grades?

Student 1:   I remember a time, but it might not have been the first time.

SK 2:   Let's take that time. Can you tell me what happened?

Student 2:   I did very well. I remember getting a red star on the top of my paper with 100, and that stands out in my memory as exciting and interesting.

SK 3:   Yes. Is it only the red star that stands out, or what happened around it?

Student 3:   (Laughter) I remember the color very very well. It was shining. I remember getting rewarded all the way around. I remember being honored by my classmates and the teacher and my parents—them making a fuss. And some of the other kids not responding so well who didn't do so well. It was mixed emotions, but generally I remember the celebration aspect.

SK 4:   You said mixed emotions. Are you able to describe them?

Student 4:   Well, at that time I was the teacher's pet and some people would say, "Aha, maybe she didn't earn it, maybe it's just because the teacher likes her so well." And some kind of stratification occurring because I was not only the teacher's pet, but I was maybe getting better grades and it created some kind of dissonance within my classmates' experience of me socially.

SK 5:   Could you describe that dissonance?

Student 5:   Well, I think there's always some kind of demarcation between students who do well and students who don't do as well, and that's determined, especially in the primary grades, by the number that you get on top of your paper.

SK 6:   Was this early in school? Was it first grade?

Student 6:   Third grade.

SK 7:   Third grade. Well, that's a long time ago. Are you able to remember what they said? Or . . .

Student 7:   No, it was more feeling . . .

SK 8:   The feeling . . .

Student 8:   Yeah, it was the feeling of, I'd put some space between me and the peer group . . .

SK 9:   Because of your good grades . . .

Student 9:   Yeah.

SK 10:   Did you try to do anything about that?

Student 10:   I didn't do so well after that. It really affected me in a large way. I wanted to be with them more than I wanted to be with the teacher, or on the teacher's good list. So it was significant.

SK 11:   It was a significant experience (the student said yes)—to you, and you got in a conflict between teacher and your peers, or you experienced it as a conflict. (Yes.) Did your parents enter into the situation?

Student 11:   Not that I recall, because it was . . . to me it was a significant alteration in how I experienced grades. To them it was maybe just a little bit less. But it was still satisfactory, still acceptable, and I was still rewarded in general terms for doing well and not failing. So that dichotomy was respected.

SK 12:   That kind of dissonance between say loyalty to your teacher and the affection of the classmates, is that a situation you have been into other times? Does it remind you of . . . other . . . ?

Student 12:   It keeps repeating itself in my life, yes. Whenever I start taking my friends or my peer group for granted, I get some kind of message saying, Huh-uh, what's more important to me? And what's more important to me is my friendships.

SK 13:   Um-hmm. That is the basic issue . . .

You mentioned several times before "rewarded"–what do you mean by "rewarded"?

Student 13:   Oh, getting to stay up to watch TV when I was in third grade, maybe; or getting to go some place or stay out later or maybe just getting ice cream, some food . . .

[The remainder of the interview, omitted here for reasons of space, went on about the importance of the student's friendships in college, and it ended in the following exchange.]

SK 26:   Okay—are there any more things you would want to say before we end the interview?

Student 26: No, I don't think so.

SK 27:   Okay, thank you very much for your cooperation.

[The interview was then discussed in class, as excerpted below.]

SK 28:   How did you experience being interviewed about it [the grades] up in front here?

Student 27: I thought it was a really good opportunity for me to explore that. I haven't even thought about it in a long time, but I knew from therapy that I've had recently that was a big time in my life when I was closer to my teacher than I was to my friends and I've had to face that a lot. It was fun for me to talk about it 'cause I'm pretty clear about what happened.

When we look at the knowledge brought forth in this short interview passage, several important aspects of the social effects of grading appear—primarily a pervasive conflict about her loyalty to her teacher or friends; being a teacher's pet getting high grades created a dissonance in her classmates' experiences of her and put a space between her and the peer group, a dissonance that kept repeating itself in her life, with her friendships being more important (Student 3–5).

The general approach of this demonstration interview is in line with a phenomenological life world interview and its mode of understanding as discussed in twelve aspects in Box 2.2. The way of producing knowledge in this interview was inspired by Rogers's client-centered questioning. Thematically, the interviewer had, in the interview guide, wanted to address meanings of grading from three theoretical positions mentioned earlier—the Rogerian, Freudian, and Skinnerian approaches (Chapter 6). Thus, when the student described "mixed emotions" (3) and "it was more the feeling" (7), the interviewer sought, in line with a Rogerian approach, to encourage further elaboration of the feeling and the mixed emotions by repeating these very words (SK 4 and 8). A Freudian approach in a broad sense was tried by asking, "Did your parents enter into the situation?" (SK 11) and, later, whether the loyalty conflict between teacher and pupils reminded her of other situations (SK 12). The student's answer confirmed that this kept repeating itself in her life, but she did not bring up family relations. The interviewer here had in mind her grade/loyalty conflict as possibly reactivating childhood conflicts of jealousy and sibling rivalry for the affection of parents. Early in the interview the student (3) had mentioned reinforcements for good grades, such as being honored by her classmates, teacher, and parents. A Skinnerian reinforcement approach was pursued by the interviewer (SK 13) in probing the meaning of the student's term "rewarded" (3 and 13). The student then told about being rewarded for good grades as a child by getting to stay up late to watch TV or by being given ice cream (13).

SETTING THE INTERVIEW STAGE

The setting of the interview stage should encourage the interviewees to describe their points of view on their lives and worlds. The first few minutes of an interview are decisive. The interviewees will want to have a grasp of the interviewer before they allow themselves to talk freely and expose their experiences and feelings to a stranger. A good contact is established by attentive listening, with the interviewer showing interest, understanding, and respect for what the subject says, and with an interviewer at ease and clear about what he or she wants to know.

The interview is introduced by a *briefing* in which the interviewer defines the situation for the subject, briefly tells about the purpose of the interview, the use of a sound recorder, and so on and asks if the subject has any questions before starting the interview. Further information can preferably wait until the interview is over. The demonstration interview about grades was introduced with a briefing about the purpose and context of the interview before (SK 0) and at the start of the interview (SK 1).

At the end of an interview there may be some tension or anxiety, as the subject has been open about personal and sometimes emotional experiences and may be wondering about the purpose and later use of the interview. There may perhaps also be feelings of emptiness; the subject has given much information about his or her life and may not have received anything in return. This being said, a common experience after research interviews is that the subjects have experienced the interview as genuinely enriching, have enjoyed talking freely with an attentive listener, and have sometimes obtained new insights into important themes of their life worlds.

The initial briefing should be followed up with a *debriefing* after the interview. The demonstration interview was thus rounded off by a debriefing before ending the interview by asking if the student had anything more to say (SK 26), and also after the interview by asking the student about her experience of the interview (SK 28), an invitation this student accepted by commenting further on the interview theme in relation to her therapy and her biography.

An interview may also be rounded off with the interviewer mentioning some of the main points that he or she has learned from the interview. The subject might then want to comment on this feedback. After this the interaction may be concluded by the interviewer saying, for example, "I have no further questions. Is there anything else you would like to bring up, or ask about, before we finish the interview?" This gives the subject an additional opportunity to deal with issues he or she has been thinking or worrying about during the interview. The debriefing is likely to continue after the sound recorder has been turned off. After a first gasp of relief, some interviewees may then bring up topics they did not feel safe raising with the sound recorder on. And the interviewer can now, insofar as the subject is interested, tell more about the purpose and design of the interview study. If new and interesting topics come up after the sound recorder has been turned off, the interviewer should consider if and how this material can be used in the further analysis and reporting of the interview. The interviewee may feel that it is part of the agreement that the postinterview conversation should not become part of the further research process, but, at the same time, it is not unusual that important things are said here, if the respondent talks more freely, throwing new light on the research interview itself. From an ethical point of view, the interviewer ought to ask the interviewee for permission to report the topics that emerge in the informal conversation after the interview.

The live interview situation, with the interviewee's voice and facial and bodily expressions accompanying the statements, provides a richer access to the subjects' meanings than the transcribed texts will do later on. It may be worthwhile for the interviewer to set aside 10 minutes or more of quiet time after each interview to reflect on what has been learned from the particular interview. These immediate impressions, based on the interviewer's empathic access to the meanings communicated in the live interview interaction, may—in the form of notes or simply recorded onto the sound recorder—provide a valuable context for the later analysis of transcripts.

If an interview is to be reported, perhaps quoted at length, then the researcher should attempt when feasible to make the social context explicit during the interview, and also the emotional tone of the interaction, so that what is said is understandable to the readers who have not witnessed the lived bodily presence of the interview situation. Much is to be learned from journalists and novelists about how to use questions and replies to also convey the setting and mood of a conversation. A heightened attention to the numerous interviews that surround us in the contemporary interview society, such as radio and television interviews, can sensitize interviewers to the dynamics of face-to-face conversations. Although few of these interviews are qualitative research interviews (some broadcasted journalistic interviews do, however, resemble research interviews), they can still alert the careful listener/viewer to the importance and effects of body language, tone of voice, types of questions, empathic versus confronting interviewer styles, and so on.

SCRIPTING THE INTERVIEW

The interview stage is usually prepared with a script. An interview guide is a script, which structures the course of the interview more or less tightly. The guide may merely contain some topics to be covered, or it can be a detailed sequence of carefully worded questions. For the semi-structured type of interview discussed here, the guide will include an outline of topics to be covered, with suggested questions. It will depend on the particular study whether the questions and their sequence are strictly predetermined and binding on the interviewers, or whether it is the interviewers' judgment and tact that decides how closely to stick to the guide and how much to follow up the interviewees' answers and the new directions they may open up.

Interviews differ in their openness of purpose; the interviewer can explain the purpose and pose direct questions from the start, or can adopt a roundabout approach, with indirect questions, and reveal the purpose only when the interview is over. The latter approach is called a funnel shaped interview—an example of which would be an interviewer interested in exploring cultural attitudes in a community who starts by asking generally about the neighborhood, goes on to ask if there are many immigrants, and ends up specifically asking about attitudes to their Muslim neighbors. The application of such indirect interview techniques needs be considered in relation to the ethical guidelines of informed consent.

An interview question can be evaluated with respect to both a thematic and a dynamic dimension: thematically with regard to producing knowledge, and dynamically with regard to the interpersonal relationship in the interview. A good interview question should contribute thematically to knowledge production and dynamically to promoting a good interview interaction.

Thematically, the questions relate to the "what" of an interview, to the theoretical conceptions of the research topic, and to the subsequent analysis of the interview. The questions will differ depending on whether the researcher is interviewing for spontaneous descriptions of the lived world, interviewing for coherent narratives, or interviewing for a conceptual analysis of the person's understanding of a topic. The more spontaneous the interview procedure, the more likely one is to obtain unprompted, lively, and unexpected answers from the interviewees. And on the other hand: The more structured the interview situation is, the easier the later conceptual structuring of the interview by analysis will be.

In line with the principle of "pushing forward," the later stage of interview analysis should be taken into account when preparing the interview questions. If the analysis will involve coding the answers, then during the interview the researcher should continually clarify the meanings of the answers with respect to the categories to be used later. If a narrative analysis will be employed, then the researcher should give the subjects ample freedom and time to unfold their own stories, and follow up with questions to shed light on the main episodes and characters in their narratives.

Dynamically, the questions pertain to the "how" of an interview; they should promote a positive interaction, keep the flow of the conversation going, and stimulate the subjects to talk about their experiences and feelings. The questions should be easy to understand, short, and devoid of academic language. A conceptually good thematic research question need not be a good dynamic interview question. Novice interviewers may be tempted to start with direct conceptual questions. The sociologist Sennett thus received the following response when as a young student he used a rather direct approach when interviewing members of the Boston elite: " 'My what, young man?' an elderly Boston matron replied when I asked her, point-blank over tea in the Somerset

Club, to describe her identity. I had just made the tyro interviewer's error of assuming that frontal attack is the best way to elicit information from others" (2004, p. 41).

When preparing an interview, it may be useful to develop two interview guides, one with the project's thematic research questions and the other with interview questions to be posed, which takes both the thematic and the dynamic dimensions into account. The researcher questions are usually formulated in a theoretical language, whereas the interviewer questions should be expressed in the everyday language of the interviewees.

Table 7.1 depicts the translation of thematic research questions in the grading study into interview questions that could provide thematic knowledge and also contribute dynamically to a natural conversational flow. The academic

Table 7.1   Research Questions and Interview Questions

Researcher Questions	Interviewer Questions
	Do you find the subjects you learn important?
Which form of learning motivation dominates in high school?	Do you find learning interesting in itself?
	What is your main purpose in going to high school?
Do the grades promote an external, instrumental motivation at the expense of an intrinsic interest motivation for learning?	Have you experienced a conflict between what you wanted to read (study) and what you had to read to obtain a good grade?
Does learning for grades socialize to working for wages?	Have you been rewarded with money for getting good grades?
	Do you see any connection between money and grades?

research questions, such as questions about intrinsic and extrinsic motivation, needed to be translated into an easygoing, colloquial form to generate spontaneous and rich descriptions. The abstract wording of the research questions would hardly lead to off-the-cuff answers from high school pupils. One research question can be investigated through several interview questions, thus obtaining rich and varied information by approaching a topic from several angles. And one interview question might provide answers to several research questions.

The roles of the "why," "what," and "how" questions differ in the case of research questions and interview questions. When designing an interview project, the "why" and "what" questions should be asked and answered before the question of "how" is posed. In the interview situation, the priority changes; here the main questions should normally be in a descriptive form: "What happened and how did it happen?" "How did you feel then?" "What did you experience?" and the like. The aim is to elicit spontaneous descriptions from the subjects rather than to get their own, more or less speculative explanations of why something took place. Many "why" questions in an interview may lead to an overreflected intellectualized interview, and perhaps also evoke memories of oral examinations. "Why" questions about the subjects' reasons for their actions may, nevertheless, be important in their own right, and when posed, should preferably be postponed until toward the end of the interview.

The question of why the subjects experience and act as they do is primarily a task for the researcher to evaluate, and the interviewer may here go beyond the subjects' self-understanding. An analogy to a doctor's diagnosis may be clarifying. The doctor does not start by asking the patient why he or she is sick, but rather asks the patient what is wrong, what he or she is feeling and what the symptoms are. On the basis of the information from the patient interview, and from other methods of investigation, the doctor then makes a diagnosis of which illness is likely. For both the doctor and the interview researcher there are cases where it is important to know the subject's own explanations of his or her condition and to ask questions about why. The primary task for both the doctor and the interviewer, however, remains that of obtaining descriptions so that they will have relevant and reliable material from which to draw their interpretations.

In addition to paying attention to the thematic and dynamic aspects of the questions, the interviewer should also try to keep in mind the later analysis, verification, and reporting of the interviews. Interviewers who know what they are asking about, and why they are asking, will attempt to clarify the meanings relevant to the project during the interview. Such attempts at disambiguation of interviewee's statements will provide a more secure ground for the later analysis. These efforts of meaning clarification during the interview may also communicate to the subject that the researcher is actually listening to, and is interested in, what he or she is saying. Ideally the testing of hypotheses and of interpretations is finished at the end of the interview, with the interviewer's hypotheses and interpretations verified, falsified, or refined.

INTERVIEWER QUESTIONS

The interviewer's questions should be brief and simple. The introductory question may concern a concrete situation; thus the demonstration interview was opened with a question of whether the student remembered the first time she ever had any grades (SK 1). The dimensions of grading introduced in her answer were then followed up in the remainder of the interview; note the parallel here to the structure of the phenomenological interview about learning presented in Chapter 2.

The interview researcher is his, or her, own research tool. The interviewer's ability to sense the immediate meaning of an answer, and the horizon of possible meanings that it opens up, is decisive. This, again, requires knowledge of and interest in the research theme and the human interaction of the interview, as well as familiarity with modes of questioning, in order for the interviewer to devote his or her attention to the interview subject and the interview topic. The questions may vary for different subjects. Kinsey and his colleagues, whose research we discussed in the previous chapter, found that open questions provided the fullest answers and that "Standardized questions do not bring standardized answers, for the same question means different things to different people. In order to have questions mean the same thing to different people, they must be modified to fit the vocabulary, the educational background, and the comprehension of each subject" (1948, p. 52; see also Shaffer and Elkins, 2005, for a brief presentation of Kinsey's mode of interviewing).

BOX 7.1   Types of Interview Questions

A. *Introductory Questions.* "Can you tell me about . . . ?"; "Do you remember an occasion when . . . ?"; "What happened in the episode you mentioned?"; and "Could you describe in as much detail as possible a situation in which learning occurred for you?" Such opening questions may yield spontaneous, rich descriptions where the subjects themselves provide what they have experienced as the main aspects of the phenomena investigated.

B. *Follow-up Questions.* The subjects' answers may be extended through the curious, persistent, and critical attitude of the interviewer. This can be done through direct questioning of what has just been said. Also a mere nod, or "mm," or just a pause can invite the subject to go on with the description. Repeating significant words of an answer can lead to further elaboration. Interviewers can train themselves to notice "red lights" in the answers—such as unusual terms, strong intonations, and the like—which may signal a whole complex of topics important to the subject.

C. *Probing Questions.* "Could you say something more about that?"; "Can you give a more detailed description of what happened?"; "Do you have further examples of this?" The interviewer here pursues the answers, probing their content, but without stating what dimensions are to be taken into account.

D. *Specifying Questions.* The interviewer may also follow up with more operationalizing questions, for instance: "What did you actually do when you felt a mounting anxiety?"; "How did your body react?" In an interview with many general statements, the interviewer can attempt to get more precise descriptions by asking, "Have you also experienced this yourself?"

E. *Direct Questions.* The interviewer here directly introduces topics and dimensions, for example: "Have you ever received money for good grades?" "When you mention competition, do you then think of a sportsmanlike or a destructive competition?" Such direct questions may preferably be postponed until the later parts of the interview, after the subjects have given their own spontaneous descriptions and thereby indicated which aspects of the phenomena are central to them.

(Continued)

(Continued)

F. *Indirect Questions.* Here the interviewer may apply projective questions such as "How do you believe other pupils regard the competition for grades?" The answer may refer directly to the attitudes of others; it may also be an indirect statement of the pupil's own attitude, which he or she does not state directly. Careful further questioning will be necessary here to interpret the answer.

G. *Structuring Questions.* The interviewer is responsible for the course of the interview and should indicate when a theme has been exhausted. The interviewer may directly and politely break off long answers that are irrelevant to the investigation, for example by briefly stating his or her understanding of an answer, and then saying, "I would now like to introduce another topic: . . ."

H. *Silence.* Rather than making the interview a cross-examination by continually firing off questions, the research interviewer can take a lead from therapists in employing silence to further the interview. By allowing pauses in the conversation the subjects have ample time to associate and reflect and then break the silence themselves with significant information.

I. *Interpreting Questions.* The degree of interpretation may involve merely rephrasing an answer, such as "You then mean that . . . ?" or an attempt at clarification, such as "Is it correct that you feel that . . . ?" or "Does the expression . . . cover what you have just expressed?" There may also be a more direct interpretation of what the pupil has said; for example, "Is it correct that your main anxiety about the grades concerns the reaction from your parents?"

Box 7.1 depicts some main types of questions that may be useful, several of which were applied in the demonstration interview. The introductory question, asking about a specific episode of grading (SK 1/Question type A), hit home, and the first two thirds of the interview were mainly a follow up (B) of the student's answer (Student 2) about the "red star." The term, and probably also her voice and facial expression, had indicated that this was a symbol of some significant experience. The interviewer's follow-up question, repeating "red star" (SK 3/B), led to an emotional response rich in information (Student 3).

Continued probing, repeating another significant expression—"mixed emotions"—and probing for further description (SK 4/B and C) opened up to a basic conflict for the subject between loyalty to the teacher or to her peers. This theme was pursued until the concluding student remark, "And what's more important to me is my friendships" (12). In some of the answers in this sequence, the interviewer overheard potentially significant expressions like "demarcation" and "space" (Student 5 and 8), and instead of following them up, posed specifying (SK 6/D) and interpreting (SK 9/I) questions.

The majority of questions in this interview were probing (C)–often by repeating significant words from the student's answers to the few direct questions about episodes and effects of grading. There were a few interpreting questions, as when early in the interview a direct interpretation of the student's statement (8) "Because of your good grades?" (SK 9/I) was immediately confirmed by a "Yeah" (Student 9). A later meaning-clarifying question "What do you mean by 'rewarded'?"(SK 13/I) led to descriptions of specific rewards, such as getting ice cream or staying up to watch TV (Student 13).

There were also answers, which were not followed up on, such as the student's emphasizing "What's most important to me is friendship" (Student 12). That is, the interviewer tried to do so by acknowledging what she said—"That is the basic issue" (SK 13), but when she did not pick it up, the topic was changed to reward for grades. As the interviewer, I (Steinar Kvale) no longer recall whether I chose not to pursue the friendship theme because I felt it was becoming too sensitive for her to elaborate further on in front of the class, or whether I sensed that she was changing my attempt to keep the interview within a discourse of grades to a discourse about friendships.

BOX 7.2   Linguistic Forms of Questions

Can you describe it to me? What happened?

What did you do? How do you remember it? How did you experience it?

What do you feel about it? How was your emotional reaction to this event?

What do you think about it? How do you conceive of this issue?

What is your opinion of what happened? How do you judge it today?

With attention to the linguistic nature of interview interaction, there follows a focus on the wording of the questions, which should be adapted to the subject matter and the purpose of an interview study. In Box 7.2 different linguistic forms of a question asking for elaboration on an episode are suggested. The wording invites rather different styles of answers, going from descriptive, behavioral, and experiential domains to emotional, cognitive, and evaluative realms. A consistent use of one type of questioning throughout an interview may lead to a specific style of answer, resulting for example in a predominantly emotional or conceptual interview.

The questions in the interview about grades were mostly matter-of-the-fact questions, asking the student to "remember" (SK 1 and 7), "what happened" (SK 2 and 3), and to "describe" (SK 4 and 5). There was a conceptual question phrased "what do you mean by" (SK 13). There were no direct questions about the feelings, but in two cases the interviewer repeated emotional terms the student used—"mixed emotions" (SK 4) and "feeling" (SK 8), and the two questions asking for descriptions concerned the student's feelings—"mixed emotions" and "dissonance" (SK 4 and 5). In contrast, in the client-centered interview in Box 2.5, the counselor predominantly used emotional terms in his questions, and Socrates applied inquisitive conceptual questions in his pursuits of the essence of justice, love, and beauty.

THE ART OF SECOND QUESTIONS

Until now the focus has been on the interviewer's questions. Now we focus on second questions, which involve active listening—the interviewer's ability to listen actively to what the interviewee says. Active listening is as important as the specific mastery of questioning techniques. The interviewer needs to learn to listen to what is said and how it is said. We may here, in line with the phenomenological approach depicted in Chapter 2, speak of the interviewer upholding an attitude of maximum openness to what appears. We may also note that in *Time* magazine's biographies of Freud and Piaget (see Chapter 1), their ability to listen was mentioned as a key trait ("Scientists and Thinkers," 1999). This concerned Freud's mastery of patient and largely silent listening, and Piaget's listening to and watching children, his ability of looking carefully at how knowledge develops in children.

   Decisions about which of the many dimensions of a subject's answer to pursue requires that the interviewer have an ear for the interview theme and a knowledge of the interview topic, a sensitivity toward the social relationship of an interview, and knowledge of what he or she wants to ask about. We may here perhaps draw an analogy to chess, where each move by the opponent changes the structure of the chessboard, and each player has to consider the multiple implications of the opponent's move before making the next move, anticipating the future moves of the opponent, and so on. What we also know from studies of expert chess players is that they rarely engage in explicit rule following (e.g., by thinking that if the queen moves like this, I ought to move my king like that), which in fact is characteristic of novice chess playing (Dreyfus & Dreyfus, 1986). Experts rely not on context-free rules but on their intuitive skills and feel for the game; this is closely related to what Aristotle described as *phronesis,* as discussed in Chapter 4 on ethics. The expert interviewer is likewise immersed in the concrete situation and is sensitive and attentive to the situational cues that will allow him or her to go on with the interview in a fruitful way that will help answer the research question, instead of focusing all attention on the interview guide, on methodological rules of interviewing, or on what question to pose next.

BOX 7.3   Second Questions

Pupil: Grades are often unjust, because very often—very often—they are only a measure of how much you talk, and how much you agree with the teacher's opinion.

Interviewer: How should that influence the grade?

Other potential interviewer responses:

   Silence . . .

   Hm, mm . . .

   "How much you talk"?

   Can you tell me more about that?

   Could you give some examples of what you are saying?

                                                                    (Continued)

(Continued)

Have you experienced this yourself?

You feel that the grades are not fair?

You find that the grades do not express your own abilities?

Can you describe more fully grades as "only a measure of how much you talk"?

Could you specify how one follows the teacher's opinion?

When you say grades depend upon how much you talk, do you then mean bluffing?

When you mention importance of following teacher's opinion, are you thinking of wheedling?

Are you sure that is correct?

Is this not only your postulate?

In Box 7.3 some of the multitude of potential interviewer responses are suggested in response to one interviewee statement about grades reported in Box 1.1. There is no one "correct" follow-up question; the options suggested in the box open up to different aspects of the answer. The potential responses are grouped, roughly, from merely indicating that the answer is heard and repeating a few words of the answer as an invitation to elaborate, to more specifically probing questions, to more or less interpreting questions, and finally to some counter-questions, the last one also being used by the interviewer in this case. In the next chapter, we will consider in greater detail the use of such confronting questions. The art of posing second questions can hardly be specified in advance, but requires a flexible on the spot follow up of the subjects' answers, with consideration of the research questions of the interview inquiry.

In the demonstration interview, the interviewer followed up the students' answers by asking for further descriptions and clarification of meanings and emotions. The interviewer found it fairly easy to follow up on the student's answers, as he was familiar with the topic and had conducted several interviews on grading in a study in Denmark (Box 6.4). In this short interview, the interviewer wanted to demonstrate a specific interviewing technique to the class and hoped that this might provide interesting new knowledge of a familiar topic.

BOOKS ON INTERVIEWING

In this chapter, we have demonstrated in some detail how to conduct a semi-structured life world interview. We will conclude the chapter by briefly mentioning other books on interviewing with in part different perspectives on interviewing than the present one.

BOX 7.4 Books on Interviewing

Seidman's (1991) *Interviewing as Qualitative Research* is informed by phenomenology and is a good example of a helpful "how to" book with a strong emphasis on techniques and concrete questions. This also applies to Spradley's (1979) classic, *The Ethnographic Interview*.

More theoretically informed books are Rubin and Rubin's (2005) *Qualitative Interviewing: The Art of Hearing Data*, with its "responsive interviewing" approach, which relies mainly on what the authors call interpretive, constructionist philosophy, mixed with critical theory (p. 30). Large parts of the book are concerned with the interview interaction itself: how to engage in a "conversational partnership," how to structure the conversation, and how to ask questions. Furthermore, the book contains concrete and helpful advice about many practicalities of the qualitative research process, for example concerning how to get one's material published. Ethical issues, wider epistemological issues, and issues about the interview as a social practice in contemporary society play only a minor role in the book.

Wengraf's (2001) *Qualitative Research Interviewing* is also about semi-structured depth interviews, but with a specific focus on biographic narrative interviewing. It is a conceptual and technical account of interviewing with many special terms that are used in precise and consistent ways. Wengraf distinguishes "receptive strategies" from "assertive strategies" in interviewing, with the former being close to Carl Rogers's model of psychotherapy (p. 154), and the latter being more in line with active and discursive approaches to interviewing (p. 155). Like most interview researchers, the author mainly sides with the former receptive strategy. Like Rubin and Rubin's book, Wengraf's book contains little on ethics and interviewing as a social practice, but it is a thorough account of biographic narrative methods.

Briggs's (1986) *Learning How to Ask* is based on an ethnographic approach and distinguishes itself by its focus on interviewing as a practice that often rests on Western conception of communication and reality, and Briggs calls for further research into interviewing itself.

EIGHT

INTERVIEW VARIATIONS

⸺◦•◆•◦⸺

In this chapter, we present a toolbox of different forms of interviewing. They are suitable for different research purposes, and the idea of this chapter is to enlarge the tool chest of the qualitative research interviewer. First we discuss some of the issues that may arise when interviewing different subjects, such as foreigners, children, and elites. Then, we outline a heterogeneous range of interview forms. In contrast to the detailed application of a phenomenological life world interview in the preceding chapter, in this chapter we present the varieties of interviewing more briefly, give examples, and refer to literature for further inspiration.

The various forms of interviews are different tools that the interviewer may choose among, depending on the purpose of the inquiry, the kind of knowledge sought, the interview subjects, and the personal skills and style of the interviewer. Our main point is that different forms of interviews are needed for different purposes, just as a craftsman needs a number of different tools in the toolbox. As a saying goes: If all we have is hammer, the whole world will look like a nail! Factual, conceptual, and focus group interviews, as well as narrative and discursive interviews, imply different social dynamics and questioning techniques. We will conclude by contrasting the more harmonious and empathic life world interviews of the preceding chapter with more agonistic and confrontational forms of interviewing.

INTERVIEW SUBJECTS

The interview form discussed in the preceding chapter pertains primarily to middle-class adult subjects in Northern Europe and America. Different issues pertain to interviewing children and elites, men and women, police suspects and witnesses. Furthermore, it requires special considerations to conduct interviewing across cultures. We will highlight some of the issues raised when interviewing different populations by focusing on cross-cultural interviews and interviews with children and elites.

Interviewing Subjects Across Cultures

Foreign cultures may involve different norms for interaction with strangers concerning initiative, directness, modes of questioning, and the like. When doing cross-cultural interviewing it is difficult to become aware of the multitude of cultural factors that affect the relationship between interviewer and interviewee. In a foreign culture, an interviewer needs time to establish familiarity with the new culture and learn some of the many verbal and non-verbal factors that may cause interviewers in a foreign culture to go amiss. For example, the simple word "yes" is in some cultures heard as an agreement, whereas in other cultures it is a response just confirming that the question has been heard, a difference that may be crucial in, for instance, negotiations of business contracts. Extralinguistic features of communication may also give rise to intercultural misunderstandings, such as when cultural groups use similar gestures, but with different meanings intended; thus nodding, which in most parts of Europe signifies agreement, in several areas of Greece means "no" (Ryen, 2002).

Some of the specific factors that may be critical in cross-cultural interviewing include asking questions as a means of obtaining information; making direct rather than circuitous replies; referring directly to matters that are taboo; looking into a person's face when speaking; sending a man to interview a woman, and vice versa (Keats, 2000). In addition to this, the linguistic and social issues of translation are important. Care should be taken to select an interpreter who is culturally acceptable as well as proficient in the language. The role of the interpreter is to assist, and not to take over the role of the interviewer or the interviewee. In particular, this may be a risk when, instead of using a professional translator, a relative or friend serves as an interpreter.

While this may facilitate contact, the familiar interpreter may have an agenda of his or her own and subtly enter into an interviewer or interviewee role.

Difficulties in recognizing disparities in language use, gestures, and cultural norms may also arise within a researcher's own culture when interviewing across gender and generation, or social class and religion. Such differences between subcultures may not be as pronounced as those between different cultures, but if the researcher makes the implicit assumption that everyone belongs to a common culture, intracultural variations may be difficult to detect.

Interviews With Children

Interviews with children allow them to give voice to their own experiences and understanding of their world. In particular, Piaget's interviews with children about their physical concepts and understanding of reality and morality have shaped our current views of children's thought processes.

BOX 8.1   Piaget's Interview About a Child's Dreams

Piaget: Where does the dream come from?

Child (5 years): I think you sleep so well that you dream.

Piaget: Does it come from us or from outside?

Child: From outside.

Piaget: What do we dream with?

Child: I don't know.

Piaget: With the hands? . . . With nothing?

Child: Yes, with nothing.

Piaget: When you are in bed and you dream, where is the dream?

Child: In my bed, under the blanket. I don't really know. If it was in my stomach, the bones would be in the way and I shouldn't see it.

Piaget: Is the dream there when you sleep?

Child: Yes, it is in the bed beside me.

SOURCE: *The Child's Conception of the World* (pp. 97–98), by J. Piaget, 1930, New York: Harcourt, Brace, and World.

In the interview passage in Box 8.1, Piaget consistently challenges the child's understanding of the location of dreams, thereby attempting to arrive at the child's conception of dreams. However, we may note that in this passage it is the adult interviewer who introduces and persists with questions of the spatial location of dreams, a dimension that does not seem to be a central issue for the child. The child appears to be influenced by the interviewer's suggestions and leading questions, thus twice repeating verbatim the interviewer's proposals as his own answers: "from outside" and "with nothing."

The influence of leading questions becomes crucial in criminal cases involving children. The reliability of information obtained through interviews with children is critical in court cases about child abuse. Children may be reluctant to talk to a stranger about painful events. Children are also easily led by the questions of adults and may provide unreliable or directly false information.

While it may be a truism that children and adults live in different social worlds, the many differences may be easy to overlook when interviewing children. Eder and Fingerson (2002) draw attention to the power imbalance between the child and the adult, and the need for the interviewer to avoid being associated with the classroom teacher, as well as to refrain from conveying expectations that there is one right answer to a question. Often, children accept questions that adults would have rejected, and attempt to answer them, such as in Piaget's interview above. An example of the second author's son, who was afraid to stay in his room because of an alleged goblin, illustrates this. Confronted with the leading question, posed by the father, whether the goblin in the room was riding a bike with one wheel or one with two wheels, the boy had to accept that it was a bike-riding goblin, which, then, could naturally not pose a threat (for bike-riding goblins can be nothing but friendly)!

It is important to use age-appropriate questions, and several difficulties of interviews with adults may be aggravated in interviews with children, such as the interviewer asking long and complex questions and posing more than one question at a time. Some barriers between children and adults may be bridged when interviewing children in natural settings. In quite a few instances interviews with children may preferably take place within the context of some other task, such as drawing, reading a story, watching TV or a video, playing with dolls or cars. Many of Piaget's interviews were carried out in relation to experimental tasks, such as the child judging the weight or size of objects.

Interviews With Elites

Elite interviews are with persons who are leaders or experts in a community, who are usually in powerful positions. Obtaining access to the interviewees is a key problem when studying elites, as discussed by Hertz and Imber (1995) in their anthology on elite interviews. When an interview is established, the prevailing power asymmetry of the interview situation may be canceled out by the powerful position of the elite interviewee.

Elites are used to being asked about their opinions and thoughts, and an interviewer with some expertise concerning the interview topic may provide for an interesting conversation partner. The interviewer should be knowledgeable about the topic of concern and master the technical language, as well as be familiar with the social situation and biography of the interviewee. An interviewer demonstrating that he or she has a sound knowledge of the interview topic will gain respect and be able to achieve an extent of symmetry in the interview relationship. See Zuckermann (1972) for an informative account of her interviews with Nobel laureates that depicts her initial contact with the scientists; the importance of her background knowledge of their fields and careers—showing that she had done her homework, which legitimized spending time on the interview; and the necessity of her being very precise in her questions to this intellectual elite. An example of more entertaining preparations for an elite interview is the first author's watching numerous soccer games, discussing soccer with friends, and reading soccer literature before interviewing the coach of the national Norwegian soccer team about his training strategies.

Experts may be used to being interviewed, and may more or less have prepared "talk tracks" to promote the viewpoints they want to communicate by means of the interview, which requires considerable skill from the interviewer to get beyond. Elite interviewees will tend to have a secure status, so it may be feasible to challenge their statements, with the provocations possibly leading to new insights. Interviews with experts, where the interviewer confronts and also contributes with his or her conceptions of the interview theme, may approximate the intense questioning of a Socratic dialogue, ideally leading to knowledge in the sense of *episteme* (see Chapter 3).

INTERVIEW FORMS

A variety of interview forms exist that are useful for different research purposes. In what follows, we present computer-assisted interviews, focus group

interviews, factual interviews, and conceptual interviews. Thereafter we treat
in greater detail narrative and discursive interviews, and we bring the chapter
to a close with a depiction of confrontational interviews, which diverge from
the more harmonious life world interview of the preceding chapter.

The following presentation of interview forms seeks to aid readers in selection of interview forms. For the actual application of the interview forms readers are recommend to take a look at more detailed presentations in other books,
such as Gubrium and Holstein's *Handbook of Interview Research* (2002). In
addition to the types of interviews that we will address in this chapter, other
specific forms exist, such as different forms of narrative interviews like the life
story interview (Atkinson, 1998) and biographic narrative interviews (Wengraf,
2001). Feminist (Reinharz & Chase, 2002) and collaborative forms of interviewing (Ellis & Berger, 2003) distinguish themselves by a particular focus on
the researchers' experience, sometimes expressed through autoethnography, an
approach that seeks to unite ethnographical and autobiographical intentions.
Other more or less distinct brands of "unstructured interviewing" mentioned by
Fontana and Prokos (2007) include oral history; creative interviewing (Douglas,
1985) that is close to oral history but used for a number of research purposes;
postmodern interviewing; "polyphonic interviewing," where participants' voices
are reported with little influence from the researcher; and finally, "empathetic
interviewing," which implies taking a stance in favor of those studied.

With the broad variety of interviews and subjects described in this chapter, it becomes understandable that there are no general standard procedures
and rules for research interviewing. The different interview forms further
relate to different kinds of knowledge. Whereas factual interviews are typically
in accord with a modernist conception of knowledge as preexisting elements
to be unearthed from the depths of the respondents, discursive and most forms
of narrative interviews are in line with a postmodern conception of knowledge
as constructed through social interaction, discourses, and narratives. In line
with the epistemological points discussed in Chapter 3, we believe that the
choice of how to do interviews is related to one's epistemological standpoint
and depends on what one wants to know about.

Computer-Assisted Interviews

The forms of interviewing that are treated in detail in this book are all face-
to-face interviews, involving a bodily presence with access to nonlinguistic

information expressed in gestures and facial expressions. Other forms of interviewing that are increasingly applied, however, are mediated by technologies such as the telephone or the computer. The analysis of telephone conversations was pioneered by conversation analysts, and the use of telephone interviews in qualitative inquiry has a number of advantages, such as increased opportunities to talk to people who are geographically distant from the researcher or who are located in dangerous places (e.g., war zones) (Elmholdt, 2006).

Recently, computer-assisted interviewing has become especially widespread (Couper & Hansen, 2002). Computer-assisted interviewing can be conducted through e-mail correspondence, implying an asynchronous interaction in time, with the interviewer writing a question and then waiting for a reply, or through chat interviews, mediated, for example, by one of the virtual communities that exist on the Internet. Chat interviews are more synchronous in time that e-mail interviews, often approaching a conversational format, with rapid turn takings, which is similar to that of face-to-face interactions. For qualitative research projects that are conducted in virtual realities on the Internet (e.g., in the form of online ethnography), computer-assisted interviews are very important, and one advantage is that they are self-transcribing in the sense that the written text itself is the medium through which researcher and respondents express themselves, and the text is thus basically ready for analysis the minute it has been typed.

The drawbacks of computer-assisted interviewing are also obvious: Both interviewer and interviewee should be relatively skilled at written communication, the mediated interaction introduces a possibly unfruitful reflective distance without cues from bodies and spoken language, and it can be difficult to generate rich and detailed descriptions (Elmholdt, 2006). With computer-assisted interviewing, it is, however, often easier than in conventional interviews to openly address intimate aspects of people's lives—an aspect that demands particular ethical sensitivity on behalf of the interviewer. On the positive side, Skårderud (2003), a psychiatrist specializing in eating disorders, has argued that computer-mediated therapy with people who have problematic relationships to their bodies may work very well, since the physical presence of a problematic body can represent an unwanted disturbance. Qualitative research interviews about such matters may also favorably be conducted through the Internet in order to avoid feelings of shame concerning the visible body. As with all forms of interviews, computer-assisted interviews have advantages and are suitable for some purposes, but will be unsuitable for other research purposes, such as where the bodily presence and the sound of the voice are crucial for the conversation.

Focus Group Interviews

While academic interviews have generally been one-to-one interviews, there is today an increasing use of focus group interviews. Social researchers employed group interviews in the 1920s, but widespread use of group interviewing first took place after the 1950s, when market researchers developed what they termed "focus group interviews" to investigate consumer motives and product preferences. Today focus groups dominate consumer research and are used for many purposes, ranging from the promotion of cereals to the marketing of politicians. They are also being applied in a variety of fields, such as health education and in evaluation of social programs, and in the 1980s they entered academic social research.

A focus group usually consists of six to ten subjects led by a moderator (Chrzanowska, 2002). It is characterized by a non-directive style of interviewing, where the prime concern is to encourage a variety of viewpoints on the topic in focus for the group. The group moderator introduces the topics for discussion and facilitates the interchange. The moderator's task is to create a permissive atmosphere for the expression of personal and conflicting viewpoints on the topics in focus. The aim of the focus group is not to reach consensus about, or solutions to, the issues discussed, but to bring forth different viewpoints on an issue. Focus group interviews are well suited for exploratory studies in a new domain, since the lively collective interaction may bring forth more spontaneous expressive and emotional views than in individual, often more cognitive, interviews. In the case of sensitive taboo topics, the group interaction may facilitate expression of viewpoints usually not accessible. The group interaction, however, reduces the moderator's control of the course of an interview, and one price of the lively interaction may be interview transcripts that are somewhat chaotic.

Factual Interviews

Qualitative interviews do not only focus on the interviewees' own perspectives and meanings. Obtaining valid factual information may be crucial in many interviews. Thus, in professional settings it may be vital for a medical doctor interviewing a child, or the child's parents, to acquire correct information about the exact bottle of medical pills the child had eaten from. In forensic interviews it can be imperative for a police officer or lawyer to gain valid information about whether the accused had a knife in his hand or not. Interviewing about a suspect's intentions with a knife involves a specific kind of questioning. In less dramatic settings, when interviewing for the oral history of a community (e.g., Bornat, 2000) the focus will be less on the storyteller's own perspective upon the events recounted than on his or her stories as avenues to reliable information about a collective past.

The intricacies of interviewing for factual information are well documented in studies of witness psychology. The importance of the wording of questions was forcefully brought out in an experiment where different groups of subjects were shown the same film of two cars colliding and afterwards asked about the speed of the cars. The average speed estimate in reply to the question "About how fast were the cars going when they contacted each other?" was 32 mph. Other subjects—seeing the same film, but with "contacted" replaced by "smashed" in the question—gave an average speed estimate of 41 mph (Loftus & Palmer, 1974). Such experiments may serve interviewers as a reminder to be extremely careful when wording their questions in interviews for factual information.

Conceptual Interviews

The purpose of an interview can be conceptual clarification. An interviewer may here want to chart the conceptual structure of a subject's or group of subjects' conceptions of phenomena such as of "fairness" and "competition," or "respect" and "responsibility." The latter concept was explored in a confrontational interview excerpt in Chapter 2 from Bellah and colleagues' study of North American character and values. The questions in conceptual interviews explore the meaning and the conceptual dimensions of central terms, as well as their positions and links within a conceptual network. Doing conceptual interviews can serve to uncover respondents' discourse models, that is, their taken-for-granted assumptions about what is typical, normal, or appropriate (Gee, 2005), and can favorably be conducted in concert with questions that ask for concrete descriptions, which sometimes give interesting points of contrast. An interview study of moral dilemmas arising in psychotherapeutic practice can illustrate this.

BOX 8.2   Ethical Concepts Versus Ethical Behavior

In an interview study of clinical psychologists' moral experiences, the respondents were asked about concrete experiences of ethical dilemmas and also, with inspiration from Socrates, about how to understand and define the concept of ethics.

Interviewer: How do you understand the words morality and ethics? What do they mean?

Psychologist: Well, morality is something invented by us humans, right. It is a kind of rules about how to relate to one another, so to say. That is what I think about it.

[. . .]

Later when describing a concrete ethical dilemma (concerning a client's romantic attraction to the psychotherapist), the psychologist continues:

Psychologist: Then, suddenly, one feels the ethical dilemma, right.

Interviewer: Do you feel it?

Psychologist: I do feel it. I feel it immediately, right. That—oops—now you have to be careful, right. What is it you are about to do in this instance?

Across the interviews more generally, morality was conceptualized reflectively with words like 'timetable,' 'set of directions,' 'basic rules,' 'rules of behavior,' and 'regulating values.' This contrasts with the practitioners' moral narratives, where they talk about 'intuition,' 'feelings,' and notably 'the stomach' as an ethical indicator, e.g. "sometimes I can feel it in my stomach what is right and wrong" and "sometimes I definitely have a feeling, it's a sensation in the stomach like: ouch! That was bad!."

SOURCE: *Psychology as a Moral Science*, by S. Brinkmann, 2006, Unpublished doctoral dissertation, University of Aarhus, Department of Psychology, Aarhus, Denmark.

The psychotherapists' statements, an excerpt of which is given in Box 8.2, testify to divergences between their conceptual and practical understandings of ethics. On a conceptual level, ethics was overwhelmingly described as "rules for behavior," whereas reference to rules was absent when concrete examples were given, and respondents talked very often about the importance of gut feelings, rather than guidelines, principles, and rules in their stories of dealing with concrete moral problems. These contradictions in the psychotherapists' statements indicate that there is a genuine dissimilarity between the theoretical and the practical understanding of morality in their practice.

The intricacies of interviewing about conceptual networks can be seen in anthropologists' studies of kinship structures in a foreign culture, where the questioning may concern finding the linguistic terms for the different types of relatives, and establish whether, for example, there are different terms for elder and younger, female and male, second cousins, as well as whether the terms depend on the gender of the person speaking. A conceptual interview may also be in the form of a joint endeavor to uncover the essential nature of a phenomenon, such as in Socrates' epistemic inquiries with Agathon and Cephalus to establish the essence of beauty and doing right, respectively.

Narrative Interviews

Narrative interviews center on the stories the subjects tell, on the plots and structures of their accounts. The stories may come up spontaneously during the interview or be elicited by the interviewer. In a study on narratives in interview research, *Research Interviewing: Context and Narrative*, Mishler (1986) outlined how interviews understood as narratives emphasize the temporal, the social, and the meaning structures of the interview. In everyday conversations, answers to questions often display the features of narratives, and Mishler posits that when stories appear so often, it supports the view that narratives are one of the natural cognitive and linguistic forms through which individuals attempt to organize and express meaning and knowledge. Box 8.3 depicts the start of a transcription of a narrative interview with a furniture craftsman-artist by Mishler. Mishler has included annotations, which keep the transcription close to the original oral style, but which may be distracting to an inexperienced reader (some of the annotations for the transcription are explained in Box 10.1).

BOX 8.3   A Craftsman's Narrative

I: And uh I'd like to start by asking you uh about the beginning/ uh how
you u:uh got into the . work that you now do. What was happening
then? a:ah What led you into it?

R: hm hmm. . . . hh well it's- it's strange

It ah- it ah- When I first started ah doing woodworking
I got into a program in- in a- a trade school/in high school level
and (sigh) it kin- I was from a working class background
so.hh the options seemed to be pretty limited to me.

But I always had an interest in building
even when I was in grammar school.
I was always building at night you=know
like making airplane models and things like that.
Those were the things tha- I-

Then I got to a point where I began to get experimental
and more interested in uh doing things on my own
so I would sort of design the airplane/ and build it you = know/and
    see if I could make it fly
that kind of thing.

But then- and so my- to continue my interests/I got into the wood-
    working program in the- in the-in the trade school
and it- I got bored stiff you = know.

Just- uh they- they took patterns down off the walls/and you=know
    it- uh build tha- that kind of thing.
I wasn't very interested in rebuilding like reproductions and things
    like that
so . . . hh I- I quit that.

SOURCE: *Storylines: Craftartist's Narratives of Identity* (pp. 73–74), by E. G. Mishler,
1999, Cambridge, MA: Harvard University Press.

The initial question about how the craftsman got into his work opened up to
a spontaneous story about the demotivating effect trade school had on him. On
the form side we may note how Mishler chose to render the transcript verbatim

and in a style close to a poem, where the poetic shape provides an accessible structure for the story.

In a narrative interview the interviewer can ask directly for stories, and perhaps together with the interviewee attempt to structure the different happenings recounted into coherent stories. The interviewer may introduce the interview with a question about a specific episode, such as "Can you please tell me the story of what happened at the demonstration when the police broke it up?" or about an institutional period, such as "Can you tell me about how you came to the hospital and what happened during your stay there?" Or the interviewer may ask for a life story, opening with "Please tell me about your life— you were born in . . . ?" After the initial request for a story, the main role of the narrative interviewer is to remain a listener, abstaining from interruptions, occasionally posing questions for clarification, and assisting the interviewee in continuing to tell his or her story. Through questions, nods, and silences, the interviewer is a co-producer of the narrative. Being familiar with narrative structures, the interviewer may take care to unfold temporal sequences, focus on who is the hero of the story and who are the antagonists and who are the hero's helpers, and try to ascertain the main plot of the story, the possible subplots, and the elements of tensions, conflicts, and resolutions.

Narrative interviews can serve multiple purposes, of which three will be pointed out here. First, a narrative can refer to a specific episode or course of action significant to the narrator, leading to a *short story*. Second, the narrative may concern the interviewee's life story as seen through the actor's own perspective, and is then called a *life history*, or biographical interview (Rosenthal, 2004). Third, there is the *oral history* interview, where the topic goes beyond the individual's history to cover communal history; here the interviewee is an informant, recording the oral history of a community (Bornat, 2000; Yow, 1994).

Discursive Interviews

Discourse analysis focuses on how knowledge and truth are created within discourses, and on the power relations of discourses. Discourse analysis was introduced by Foucault (e.g., 1972) and has since been developed in a variety of forms. Discourse analysis is "the study of how talk and texts are used to perform actions" (Potter, 2003, p. 73). Thus, it studies how individuals and groups utilize language to enact specific activities and identities (Gee, 2005). All interviews are naturally discursive and imply different discourses.

Interviewers working within a discursive framework will, however, be particularly attentive to specific aspects of the interaction of the interview discourse, which differ from conventional interviewing:

> First, variation in response is as important as consistency. Second, techniques, which allow diversity rather than those which eliminate it are emphasized, resulting in more informal conversational exchanges and third, interviewers are seen as active participants rather than like speaking questionnaires. (Potter & Wetherell 1987, p. 165)

A discursive perspective sensitizes the interviewer to differences in the discourses of the researcher and the subjects during an interview, and their differential powers to define the discourses. A discursive interviewer will be attentive to, and in some cases stimulate, confrontations between the different discourses in play. During the grade interview presented in the previous chapter, the interviewer was, however, not aware of how the student attempted to turn the discourse on grades, intended by the researcher, into a discourse about friendships.

BOX 8.4   Crossing Interview Discourses

I: One thing I have been wondering. A lot of you guys stay after work to do troubleshooting on your own equipment or moonlighting?

B: Yes.

I: What do you learn from that?

B: It depends on what kind of moonlighting we do. Of course it is just routine, our interest to be allowed to potter around with something in which we see some benefit. If you have an old computer monitor at home and it's broken then you bring it and fiddle with it to see if you can find the trouble. It's not . . . you know, we are not allowed to potter with our television at work. You do not learn that, you learn about an instrument. To build your own amplifier is also something else than measuring some electronic equipment down here.

I: Okay, so you do it to get some experience with more types of instruments and equipment?

B: No, it's not to become experienced, it's done to apply what you have learned at school for your own profit. Such a computer monitor—you get a new one without having to buy one. If you build an amplifier, well, then you might, it's much cheaper than having to buy one yourself. It's not to learn something extra; it's done simply out of simple interest. Or because there is some cool cash involved in repairing the video of a friend.

—

I: Okay. You express sort of a contradiction when you say it's not to learn something, it's just your interest?

B: You don't think of it as learning.

I: But you learn something through it?

B: Yes, but it's not like when you come home from school and say, 'I don't understand this, now I want to learn.' And then you go and ask for a task where it's involved. It's not like you go and choose a monitor to learn about it. You have a monitor at home which is broken and you decide to fix it. Then you find out something about it.

SOURCE: "The Research Interview as Discourses Crossing Swords," by L. Tanggaard, 2007, *Qualitative Inquiry, 13,* pp. 169–170.

In the crossing discourses of Box 8.4, the interviewer is clearly not sensitive to what a vocational pupil is trying to tell her about his work. The researcher and pupil talk in crossing discourses regarding which of his activities were to be considered learning. The researcher conceived of learning as embedded in everyday activities, whereas it is clear from the whole interview that learning was something that took place in a school. In contrast to many interviewees, this interviewee does not respond to the power position of the researcher, but takes a tough position against the researcher about whether his activities constitute learning or not. The researcher was investigating the learning of electronics at a vocational school and she had obvious problems getting this pupil to talk about learning—the very topic of her Ph.D. project. During the interview, she was not aware of how she and the pupil were following crossing discourses about what constituted learning, discourses that hardly touched each other. When analyzing the interview transcript later, after having read Foucault on discourse analysis, Tanggaard came to regard this sequence, and also her interviews with other pupils, as "discourses crossing swords," yielding important information about which activities the pupil considered learning and which not.

Some discourse analysts working with conversation analysis believe that interviews are overused in qualitative inquiry (e.g., Potter & Hepburn, 2005), and prefer to study naturally occurring speech and action. Qualitative interviews cannot be used for all research purposes, but from a discourse theoretical point of view, focus-group interviews in particular are capable of bringing into play important discourses that people use to establish social bonds and identities. In a discussion of "discourse-analytic interviewing," Parker recommends seeing the interviewee as a co-researcher in the form of a fellow discourse analyst. A discursive approach to interviewing highlights the local and dynamic meaning production of the interview situation itself in contrast to an idea of fixity of meaning: "We are taking part in a conversation with someone rather than sitting puzzling ourselves over some fixed text" (Parker, 2005, pp. 95–96).

CONFRONTATIONAL INTERVIEWS

Active and confrontational interviewing contrasts with the prevailing forms of empathic and consensus-seeking interviewing. In *The Active Interview,* Holstein and Gubrium (1995) developed the constructionist idea that all knowledge is created from the action taken to obtain it. This led them to portray interviews as unavoidably interpretively active, meaning-making practices. Traditional perspectives on the research interview recommended non-directional, unbiased questioning, given the presupposition that the subject was "epistemologically passive, not engaged in the production of knowledge." (p. 8). Holstein and Gubrium wanted to show that *all* interview responses are products of interpretive practice. In their active approach to interviewing, a basic assumption became that interview research *produces* meaning rather than *uncovers* antecedent meaning elements from the depths of the respondent's self. In active interviews, the interviewer activates narrative production, suggests narrative positions, resources, and orientations. Provoked by the interactional and informational challenges of the interview situation, the respondent becomes a kind of researcher in his or her own right, actively composing meaning by way of situated, assisted inquiry, where the interaction between interviewer and respondent is a reality-constructing process.

Holstein and Gubrium's approach to the active interview is primarily centered on an epistemological level rather than the level of concrete interaction and how-to-do interviewing. Concerning the concrete interactional level, we believe that a direct confrontational approach can bring the conflict and power dimensions of the interview conversation into focus. Such an agonistic understanding of the conversation is at the root of Lyotard's analysis of knowledge in the postmodern society. He regards every statement as a move in a game, which is "underlying our method as a whole: to speak is to fight, in the sense of playing, and speech acts fall within the domain of a general agonistics." (1984, p. 10). Agonistic interviewing enhances confrontation, where the interviewer deliberately provokes conflicts and divergences of interests, as seen in some forms of journalistic interviews. In contrast to an ideal of consensus and harmony, the interview becomes a battleground where the interviewer contradicts and challenges the interviewee's statements, whereby conflicts and power become visible. The position of the interviewer comes more into the open in a confrontational interview, with options for subjects to challenge the interviewer's assumptions, approximating a more equal power balance of the interview interaction. The goal of the confrontational interview can be to lead to insight through dialectical development of opposites, such as in Socrates' dialectical and agonistic questioning of the Sophists, leading to thoroughly tested epistemic knowledge.

In Chapter 2, we discussed an interview sequence from Bellah and colleagues' *Habits of the Heart,* where the interviewer tried to discover at what point the subject would take responsibility for another human being. When the respondent is asked about responsibility for her own children, her initial claim that she is only responsible for her own acts is challenged. Here we see a clear confrontational stance with explicit inspiration from Socrates, described as follows: "Though we did not seek to impose our ideas on those with whom we talked, . . . we did attempt to uncover assumptions, to make explicit what the person we were talking to might have left implicit" (Bellah et al., 1985, p. 304).

Bourdieu and his colleagues (1999) likewise confronted the subjects in their study of social suffering among the downtrodden in France. In the interview sequence in Box 1.3, Bourdieu's compassion for the plight of the young men did not stop him from posing inquisitive questions and proposing conflicting interpretations of their accounts. This included direct confrontations, such as "You are not telling the whole story" and leading questions pointing to information suspected withheld, such as "What were you doing, bugging him?" In an outline of his interview approach, Bourdieu compares his interviewing to Socrates' questioning:

The "Socratic" work of aiding explanations aims to propose and not to impose. To formulate suggestions sometimes explicitly presented as such ("you don't mean that . . . ") and intended to offer multiple, open-ended continuations to the interviewee's argument, to their hesitations or searching for appropriate expression. (Bourdieu et al., 1999, pp. 614–615)

The utilization of confrontational interview forms depends upon the subjects interviewed; for some subjects, strong challenges to their basic beliefs may be an ethical transgression, while confident respondents, such as elite interviewees, may be stimulated by the intellectual challenges. A confrontational interview may thus approximate a mutual and egalitarian relationship where both parts pose questions and give answers, with reciprocal criticism of what the other says. The research interview may then become a conversation, which stimulates interviewee and interviewer to formulate their ideas about the research topics, to learn and to increase their knowledge of the subject matter of inquiry.

NINE

INTERVIEW QUALITY

———— ◈ ————

W e now turn to quality criteria for good interviewing practice. Here we focus mainly on the semi-structured life world interview that was described in Chapter 7, with the caveat that the varieties of interviewing discussed in Chapter 8 may involve different quality criteria. We begin with an interview by Hamlet to illustrate how the judgment of the quality of an interview depends on its purpose and content. Thereafter internal criteria for a good and ideal interview are suggested as well as standards for the craftsmanship of interviewing. Epistemological issues pertaining to the quality of interview-produced knowledge are treated in relation to some standard external objections to interview quality and exemplified by the question of leading questions. We conclude the chapter by pointing out how methodological and ethical criteria of good interviewing in some cases may be at odds with each other.

HAMLET'S INTERVIEW

A dramatic case may exemplify how the appraisal of an interview technique depends on the content and the purpose of the interview.

Hamlet: Do you see yonder cloud that's almost in shape of a camel?

Polonius: By th' mass, and 'tis like a camel indeed.

Hamlet: Methinks it is like a weasel.

Polonius: It is back'd like a weasel.

Hamlet: Or like a whale?

Polonius: Very like a whale.

Hamlet: (Aside) They fool me to the top of my bent. (*Hamlet,* act III, scene 2)

Our first comment on the quality of this interview concerns its length. Hamlet's interview is brief. The seven lines are, however, dense and rich enough to instigate more lengthy comments. In contrast, current research interviews are often too long and filled with idle chatter. If one knows what to ask for, why one is asking, and how to ask, one can conduct short interviews that are rich in meaning.

The quality of Hamlet's interview technique depends on how the purpose of the interview is understood. This short passage gives rise to several interpretations. At first glance the interview is an example of an unreliable technique—by using three leading questions Hamlet leads Polonius to give three entirely different answers. Thus the interview does not yield any reproducible, reliable knowledge about the *shape of the cloud* in question.

At second glance, the topic of the interview might change: The figure in question is no longer the cloud, but the *personality of Polonius* and his trustworthiness. The interview then provides reliable, thrice-checked knowledge about Polonius as an unreliable person—his three different answers are all led by Hamlet's questions. With the change in the purpose and the topic of the interview, the leading questions do not produce entirely unreliable knowledge, but involve an indirect, reliable, interview technique.

The contents of Hamlet's interview then approximate a threefold ideal of being interpreted, validated, and reported by the end of the interview. By repeating the question in different versions and each time getting the "same" indirect answer about Polonius's trustworthiness, the inter-view is "self-interpreted" before Hamlet closes off with his aside interpretation: "They fool me to the top of my bent." As to the second requirement—validation—few interview researchers today repeat so consistently as Hamlet a question in different versions to test the reliability of their subject's answers. Regarding the third requirement—reporting—the short interview has been carried out so well that it speaks for itself. We would think that, when watching the play, the audience would generally experience a Gestalt switch from the shape of the cloud to the trustworthiness of Polonius as the interview topic, even before Hamlet gives his aside conclusion.

So far, we have discussed Hamlet's interview in isolation from its context, its position in the broader drama. At a third glance, the interview appears as a display of the *power relations* at a royal court. The prince demonstrates his power to make a courtier say anything he wants. Or, the courtier demonstrates his mode of managing the power relations at the court. In an earlier scene in the play, Polonius himself gave a lesson in what in current textbooks of method is called an indirect, funnel-shaped interview technique. Polonius requests a messenger to go to Paris to inquire into the behavior of his son studying music in the city. The messenger is instructed to start with a broad approach, "Enquire me first what Danskers are in Paris," and then gradually to advance the subject, ending up with suggesting such vices as drinking, quarreling, and visiting brothels, where "Your bait of falsehood take this carp of truth," concluding the lesson, "By indirections find directions out" (*Hamlet*, act II, scene 1). With Polonius that well versed in indirect questioning techniques, is he actually caught by Hamlet's questioning technique? Or does he see through the scheme and play up to Hamlet as a courtier?

A central theme of the play, which was written at the transition from the medieval to the modern age, is a *questioning of reality*—not just a suspicion of the motives of others, but also a preoccupation with the frail nature of reality. Hamlet's interview may in that case be seen as an illustration of a *pervasive doubt about the appearance of the world,* including the shape of a cloud and the personalities of fellow players.

From an ethical perspective, the evaluation of Hamlet's interview also depends on the interpretation of its purpose and content. In the first reading, the leading questions merely lead to unreliable knowledge of the shape of the cloud. In the second reading, the interview entails the deliberate deception of Polonius; there is no question of informed consent, and the consequences may be a matter of life and death for the protagonists of the drama. Here an ethics of principles is overruled by a utilitarian interest in survival.

The quality and the ethicality of Hamlet's interview depend on the interpretation of the content and the purpose of this specific interview. With the different topics and objectives of interviews, and the variety of forms in mind, we shall nevertheless suggest some criteria for evaluating the quality of a research interview and the craftsmanship of the interviewer.

INTERVIEW QUALITY

The quality of the original interviews is decisive for the quality of the sub-sequent analysis, verification, and reporting of the interview findings, as can be seen in the emphasis by Kinsey and his colleagues on the need for developing the quality of the interaction in the interview (see the conclusion of Chapter 5 and Box 6.6). Sophisticated statistical or theoretical analysis based upon interviews of dubious quality may turn out to be magnificent edifices built on sand.

BOX 9.1   Quality Criteria for an Interview

- The extent of spontaneous, rich, specific, and relevant answers from the interviewee
- The extent of short interviewer questions and longer interviewee answers
- The degree to which the interviewer follows up and clarifies the meanings of the relevant aspects of the answers
- To a large extent, the interview being interpreted throughout the interview
- The interviewer attempting to verify his or her interpretations of the subject's answers over the course of the interview
- The interview being "self-reported," a self-reliant story that hardly requires additional explanations

---

Of the six quality criteria for a semi-structured interview proposed in Box 9.1, the last three, in particular, refer to an ideal interview—suggesting that the meaning of what is said is interpreted, verified, and reported by the time the sound recorder is turned off. This demands craftsmanship and expertise and presupposes that the interviewer knows what he or she is interviewing about, as well as why and how. Although such quality criteria might seem to be unattainable ideals, they can serve as guidelines for good interview practice. The examples given in this book of the interviews by Socrates and Hamlet do fulfill many such ideal criteria; they provide a coherent unity in themselves and present rich texts for further interpretations. Again, it should be borne in mind that different ways of doing interviews can involve different quality criteria; from the discursive viewpoint outlined in Chapters 8 and 13, the quantity of contradictions present in an interview text, for example, could be taken as a criterion that the text is potentially interesting to analyze.

THE INTERVIEW SUBJECT

Some interview subjects may appear to be better than others. Good interviewees are cooperative and well motivated; they are eloquent and knowledgeable. They are truthful and consistent; they give concise and precise answers to the interviewer's questions; they provide coherent accounts and do not continually contradict themselves; they stick to the interview topic and do not repeatedly wander off. Good subjects can give long and lively descriptions of their life situation, they tell capturing stories well suited for reporting. The subject of the interview on grades in Chapter 7 was a good interview subject according to most of these criteria.

As pleasant as such interview subjects may appear to the interviewer, it is by no means given that they provide the most valuable knowledge about the research topics in question. The idealized interviewee appears rather similar to an upper-middle-class intellectual, whose views are not necessarily representative of the general population. Well-polished eloquence and coherence may in some instances gloss over more contradictory relations to the research themes.

The ideal interview subject does not exist—different persons are suitable for different types of interviews, such as for providing accurate witness observations, versus giving sensitive accounts of personal experiences and emotional states, versus telling captivating stories. The two young men interviewed by Bourdieu (Box 1.3) were obviously not well behaved, accommodating interviewees. Still, the interview provided a strong picture of their living situation. Interviewees can be good subjects with respect to different purposes, as exemplified by Agathon providing logical contradictions for Socrates to clarify (Box 2.4) and the therapeutic client living out, and learning from, the emotional nature of the therapeutic relationship (Box 2.5). Recognizing that some people may be harder to interview than others, it remains the task of the interviewer to motivate and facilitate the subjects' accounts and to obtain interviews rich in knowledge from virtually every subject.

INTERVIEWER QUALIFICATIONS

The interviewer is the key research instrument of an interview inquiry. As an able craftsman, he or she knows how to apply interview techniques and exercise a situated judgment of the differing forms of questioning in a given interview situation. A good interviewer knows the topic of the interview, masters conversational skills, and is proficient in language, with an ear for his or her subjects' linguistic style. The interviewer should have a sense for good stories and be able to assist the subjects in the unfolding of their narratives. The interviewer must continually make on-the-spot-decisions about what to ask and how, which aspects of a subject's answer to follow up—and which not; which answers to comment on and interpret—and which not.

BOX 9.2   The Interviewer Craftsman

The interviewer craftsman is:

Knowledgeable. He or she has an extensive knowledge of the interview theme and can conduct an informed conversation about the topic. This interviewer will know what issues are important to pursue, without attempting to shine with his or her extensive knowledge.

Structuring. The interviewer introduces a purpose for the interview, outlines the procedure in passing and rounds off the interview by, for example, briefly telling what was learned in the course of the conversation and asking whether the interviewee has any questions concerning the situation.

Clear. He or she poses clear, simple, easy, and short questions; speaks distinctly and understandably; does not use academic language or professional jargon. The exception is in a stress interview: Then the questions can be complex and ambiguous, with the subjects' answers revealing their reactions to stress.

Gentle. The interviewer allows subjects to finish what they are saying, lets them proceed at their own rate of thinking and speaking. He or she is easygoing, tolerates pauses, indicates that it is acceptable to put forward unconventional and provocative opinions and to treat emotional issues.

Sensitive. He or she listens actively to the content of what is said, hears the many nuances of meaning in an answer and seeks to get the nuances of meaning described more fully. The interviewer is empathetic, listens to the emotional message in what is said, not only hearing what is said but also how it is said, and notices as well what is not said. The interviewer feels when a topic is too emotional to pursue in the interview.

Open. The interviewer hears which aspects of the interview topic are important for the interviewee, listens with an evenly hovering attention, and is open to new aspects that can be introduced by the interviewee and follows them up.

Steering. The interviewer knows what he or she wants to find out: is familiar with the purpose of the interview, what it is important to acquire knowledge about. The interviewer controls the course of the interview and is not afraid of interrupting digressions from the interviewee.

Critical. He or she does not take everything that is said at face value, but questions critically to test the reliability and validity of what the interviewees tell. This critical checking can pertain to the observational evidence of the interviewees' statements as well as to their logical consistency.

Remembering. The interviewer retains what a subject has said during the interview, can recall earlier statements and ask to have them elaborated, and can relate what has been said during different parts of the interview to each other.

Interpreting. He or she manages throughout the interview to clarify and extend the meanings of the interviewee's statements, providing interpretations of what is said, which may then be disconfirmed or confirmed by the interviewee.

Interviewer qualifications, such as those outlined in Box 9.2, which were formulated in relation to a phenomenological life world interview, may lead to good interviews in the sense of producing rich knowledge and ethically creating a beneficial situation for the subjects. In contrast, in more actively confronting interviews, as discussed in Chapter 8, the interviewer actively confronts the interviewee (e.g., by challenging him or her to provide justification for attitudes or beliefs), and the interview quality depends on the interviewer's skills in this regard. It may be more demanding for an interviewer to engage in such actively confronting interviews, especially if they involve dialectical questioning with the goal of producing knowledge in the sense of *episteme.* Such interviewing will often presuppose considerable knowledge of the subject matter under consideration. With extensive practice in different interview forms and with different subjects, an experienced interviewer might go beyond technical recommendations and criteria, and—sometimes—deliberately disregard or break the rules. Also, interviews conducted by less experienced interviewers, which do not fulfill common interview guidelines, may in some cases provide worthwhile information. The qualifications summed up here may differ for various types of interviews, and, in interviews in which the topic really matters, the technical rules and criteria may lose relevance when compared with the existential importance of the interview topic.

STANDARD OBJECTIONS TO THE QUALITY OF INTERVIEW RESEARCH

The issue of interview quality goes beyond the craftsmanship of the individual interviewer and raises epistemological and ethical issues of pursuing interview knowledge. We shall now first turn from the above internal quality criteria of interview research to some common external criticisms of the quality of interview-produced knowledge.

BOX 9.3   Standard Criticisms of Qualitative Research Interviews

The qualitative research interview is *not:*

1. scientific, but only reflects common sense

2. quantitative, but only qualitative

3. objective, but subjective

4. scientific hypothesis testing, but only explorative

5. a scientific method, since it is too person dependent

6. trustworthy, but biased

7. reliable, since it rests upon leading questions

8. intersubjective, since different readers find different meanings

9. valid, as it relies on subjective impressions

10. generalizable, because there are too few subjects

Interview reports have tended to evoke rather standardized objections about their quality from the mainstream of modern social science. In Box 9.3 10 typical criticisms of interview research are listed; the first five refer to general conceptions of scientific research, the next three to the interviewing and analysis stages, and the last two to validation and generalization. Some of the objections refer to intrinsic problems of interview research, whereas others arise from an inadequate understanding of the use of conversations as a research method. Below we give some rhetorical suggestions for responding to such standard objections, summarizing points made earlier and anticipating arguments in the coming chapters. This overview may save novice interview researchers some of the time and energy often used for external defense, to the benefit of more intensive internal work with interview quality. If an objection is considered valid to the specific interview investigation, it can be taken into account when designing the study and thereby improve the quality of the research. If an objection is regarded as invalid, the arguments for this can be offered in the report.

1. The qualitative research interview is not scientific, but only reflects common sense. No single authoritative definition of science exists, according to which the interview can be unequivocally categorized as scientific or unscientific. A working definition of science may be the methodical production of new, systematic knowledge. The question of scientific or not then depends on the understanding of the key terms in this definition, such as *methodical, new, systematic,* and *knowledge* in relation to the specific interview investigation.

2. Interviews are not quantitative, only qualitative, and thus not scientific. In paradigmatic social science discussions, science has often been equated with quantification. In the research practice of the natural and the social sciences, however, qualitative analyses also have a major position (Chapter 17). In mainstream social science textbooks on method, however, the qualitative aspects of the research process have until recently hardly existed.

3. The qualitative research interview is not objective, but subjective. The basic terms of this objection are ambiguous. The objectivity of interview research needs to be discussed specifically for each of the multiple meanings of objectivity, as relevant to the interview inquiry in question (Chapter 15).

4. Qualitative interviews do not test hypotheses; they are only explorative and thus not scientific. In a broad conception of science as hypothesis testing, as well as descriptive and exploratory, designs are important, with description and exploration as strong points of qualitative research. And, contrary to the objection, an interview may also take the form of a process of continual hypothesis testing, where the interviewer tests hypotheses, for instance with the interplay of direct questions, counter-questions, leading questions, and probing questions (Chapter 7).

5. The interview is not a scientific method, it is too person dependent. A research interview is flexible, context sensitive, and dependent on the personal interrelationship of the interviewer and interviewee. Rather than attempt to eliminate the influence of the personal interaction of interviewer and interviewee, we might regard the person of the interviewer as the primary research instrument for obtaining knowledge, which puts strong demands on the quality of the interviewer's craftsmanship (Chapter 5).

6. Interview results are not trustworthy; they are biased. The answer needs to be concrete—the specific counter-question concerns who cannot be trusted and in what sense. Unacknowledged bias may entirely invalidate the results of an interview inquiry. A recognized bias or subjective perspective, may, however, come to highlight specific aspects of the phenomena investigated and bring new dimensions forward, contributing to a multiperspectival construction of knowledge.

7. Might not the interview results be due to leading questions, and thus unreliable? The leading effects of leading questions are well documented. The qualitative interview is, however, well suited to systematic application of leading questions to check the reliability of the interviewees' answers, which was exemplified in Hamlet's interview above (more on this below).

8. The interpretation of interviews is not intersubjective, but subjective, as different readers find different meanings. We may here distinguish between an unacknowledged biased subjectivity, to be avoided, and a perspectival subjectivity. With a clarification of the perspectives adopted toward an interview text, several interpretations of the same text need not be a weakness, but can be a strong point of interview research (Chapter 12).

9. Interviewing is not a valid method, as it depends on subjective impressions. Interviewing is a personal craft, the quality of which depends on the craftsmanship of the researcher. Here validation becomes a matter of the researcher's ability to continually check, question, and theoretically interpret the findings (Chapter 15).

10. Interview findings are not generalizable; there are too few subjects. The number of subjects necessary depends on the purpose of the study. In postmodern conceptions of social sciences the goal of global generalization is replaced by a transferability of knowledge from one situation to another, taking into account the contextuality and heterogeneity of social knowledge (see also the discussion of when and how to generalize from single case studies in Chapter 15).

In a reinterpretation, the standard objections can be reversed and read as pointing to the strong points of qualitative interview research. The force of the interview is its privileged access to the subjects' everyday world. The deliberate use of the subjective perspective need not be a negative bias; rather, the personal perspectives of interviewees and interviewer can provide a distinctive and receptive understanding of the everyday life world. A controlled use of leading questions may lead to well-controlled knowledge. A plurality of interpretations enriches the meanings of the everyday world, and the researcher as a person is the most sensitive instrument available to investigate human meanings. The explorative potentialities of the interview can open to qualitative descriptions of new phenomena. Validating and generalizing from interview findings open up alternative modes of evaluating the quality and objectivity of qualitative research.

LEADING QUESTIONS

The question most likely to be asked about interview quality concerns leading questions, sometimes raised in the form of a question such as: "Cannot the interview results be due to leading questions?" The very form of the question involves a liar's paradox—an answer of "Yes, this is a serious danger" may be due to the suggestive formulation of the question leading to this answer. And a "No, this is not the case" may demonstrate that leading questions are not that powerful.

It is a well-documented finding that even a slight rewording of a question in a questionnaire or in the interrogation of eyewitnesses may influence the answer. When the results of public opinion polls are published, the proponents of a political party receiving low support are usually quick to find biases in the wording of the poll's questions. Politicians are experienced in warding off leading questions from reporters, but if leading questions are inadvertently posed to subjects who are easily suggestible, such as small children, the validity of their answers may be jeopardized.

Although the wording of a question can inadvertently shape the content of an answer, it is often overlooked that deliberately leading questions are necessary parts of many questioning procedures, as exemplified by Hamlet's interview. The validity of leading questions depends on the topic and purpose of the investigation. Legal interrogators may on purpose pose leading questions in order to obtain information they suspect is being withheld. The burden of denial is then put on the subject, as with the question, "When did you stop beating your wife?" Police officers and lawyers also intentionally apply leading questions to test the consistency and reliability of a person's statements. Piaget used questions leading in wrong directions in order to test the strength of the child's concept of, for example, weight. We may also recall Bourdieu's use of leading questions in his active confrontational interview with the two young men, such as "What were you doing, bugging him?" (Box 1.3). In Socrates' dialogue on love, he repeatedly employed leading questions with the intention of exposing the contradictions of Agathon's understanding of love and beauty and of leading Agathon to true insight.

In contrast to common opinion, the qualitative research interview is particularly well suited for employing leading questions to repeatedly check the reliability of the interviewees' answers, as well as to verify the interviewer's interpretations. Thus leading questions need not reduce the reliability of interviews, but may enhance it; rather than being used too much, deliberately leading questions are today probably applied too little in qualitative research interviews.

It should be noted that not only may the questions preceding an answer be leading, but the interviewer's own bodily and verbal responses, such as second questions following an answer, can act as positive or negative reinforcers for the answer given and thereby influence the subject's answers to further questions. We may also note that in questionnaires, the response alternatives may lead the subjects to accept the researcher's dichotomies when closing off the responses to answer either "yes" or "no" to a question, without an option to argue that a question may be based upon a false dichotomy. Leading questions may close off the range of potential answers; exemplified by the question "Which hand do you choose?" excluding answers where the subject does not want to choose either hand. Or, as an American president proclaimed not long ago: "Every nation, in every region, now has a decision to make. Either you are with us, or you are with the terrorists," which leaves those bewildered who were neither with the terrorists nor with the president. An advantage of the qualitative research interview is that the interviewee has an open range of response possibilities, including a rejection of the premises of the interviewer's questions.

While the technical issue of using leading questions in an interview has been rather overemphasized, the leading effects of project-based research questions have received less attention. Recall the different kinds of answers obtained by a Rogerian, a Freudian, and a Skinnerian approach in the imaginary interview on teasing and in the interview about grades (Chapters 6 and 7). A project's orienting research questions determine what kind of answers may be obtained. The task is, again, not to avoid leading research questions, but, in line with a hermeneutical emphasis on the role of preconceptions, to recognize the primacy of the question and attempt to make the orienting questions explicit, thereby providing the reader of an interview report with an opportunity to evaluate their influence on the research findings and to assess the validity of the findings.

The fact that the issue of leading questions has received so much attention may be due to prevailing empiricist and positivist conceptions of knowledge. There may be a belief in a neutral observational access to a social reality of objective facts independent of the investigator, implying that an interviewer collects verbal responses as a miner finds buried metals or a botanist collects plants in nature. In an alternative view of the interviewer as traveler, which follows from a postmodern perspective on knowledge construction, the interview is a conversation in which the knowledge is constructed in and through an interpersonal relationship, co-authored and co-produced by interviewer and interviewee. The decisive issue is not whether to lead or not to lead, but where the interview questions lead, whether they lead to new, trustworthy, and worthwhile knowledge.

TENSIONS BETWEEN SCIENTIFIC AND ETHICAL RESPONSIBILITY

The search for interview knowledge of high scientific quality, with the interviewees' answers critically probed and alternative interpretations checked out, may in some cases conflict with the ethical concern of not harming the interviewee

(Brinkmann & Kvale, 2005). The dilemma of wanting as much knowledge as possible, while at the same time respecting the integrity of the interview subjects, is not easily solved.

Jette Fog (2004) has formulated the interviewer's ethical dilemma: The researcher wants the interview to be as deep and probing as possible, with the risk of trespassing the person, and on the other hand to be as respectful to the interviewed person as possible, with the risk of getting empirical material that only scratches the surface. One of Fog's examples runs as follows: In a study of living with cancer, a woman is interviewed and denies that she fears a return of the disease. She says that she is not afraid, and she appears happy and reasonable. However, as a skilled interviewer and therapist, Fog senses small signals to the contrary: The woman speaks very fast, her smile and the way she moves her hands are independent of her words. Her body is rigid and she does not listen to her own words. If the interviewer decides to respect the interviewee's words, and refrains from anything resembling critical therapeutic interpretations, then the written interview will subsequently tell the story of a woman living peacefully with cancer. In this way valuable knowledge might be lost, which could only have been obtained by trying to get behind the apparent denial and defenses of the interviewee. If society has an interest in finding out what it means to live with a deadly disease, then the researcher should perhaps try to go behind the face value of the woman's words. But what is in the interest of the woman? Perhaps it is best for her not to have her defenses broken down, or maybe she will live a better life if she faces up to the reality of her disease.

Such dilemmas of conflicting scientific and ethical concerns cannot, as we argued in Chapter 4, be solved by ethical rules alone, but will depend on the ethical experience and judgment of the researcher. In some cases, research options may exist where the described dilemma does not arise. If the research interview above had been a therapeutic interview, it would have been part of the therapeutic process to go beyond the subject's apparent denials, and possibly open to hurtful self-confrontations, and as a side-effect obtain more thoroughly checked and penetrating knowledge than is ethically defensible in a research interview.

****

We will conclude that there are no unequivocal quality criteria for research interviews. A good interview rests upon the craftsmanship of the researcher, which goes beyond a mastery of questioning techniques to encompass knowledge of the research topic, sensitivity to the social relation of interviewer and interviewee, and an awareness of epistemological and ethical aspects of research interviewing. There are no fixed criteria for what constitutes a good interview, not when it comes to the scientific or the ethical quality: The evaluation of interview quality depends on the specific form, topic, and purpose of the interview.

TEN

TRANSCRIBING INTERVIEWS

W e now proceed from the live interaction of the interview situation to the postinterview stages of working with the outcome of the interview: transcribing, analyzing, verifying, and reporting the knowledge produced in the interview conversations. In this chapter we treat the transformation of the oral interview conversation to a written text in the form of transcripts amenable to analysis. First, we point out some principal differences between oral and written language. Thereafter we address the practical issues of recording and transcribing, and we raise some questions concerning the reliability and validity of transcriptions.

ORAL AND WRITTEN LANGUAGE

The quality of *interviewing* is often discussed, whereas the quality of *transcription* is seldom addressed in qualitative research literature. This may be related to a traditional lack of attention among social scientists to the linguistic medium they work with. Rather than being a simple clerical task, transcription is an interpretative process, where the differences between oral speech and written texts give rise to a series of practical and principal issues.

By neglecting issues of transcription, the interview researcher's road to hell becomes paved with transcripts. The interview is an evolving face-to-face conversation between two persons; in a transcription, the conversational interaction between two physically present persons becomes abstracted and fixed in a written

177

form. Once the interview transcriptions have been made, they tend to be regarded as *the* solid rock-bottom empirical data of an interview project. In contrast, from a linguistic perspective the transcriptions are translations from an oral language to a written language, where the constructions on the way involve a series of judgments and decisions. A transcript is a translation from one narrative mode—oral discourse—into another narrative mode—written discourse. Oral speech and written texts entail different language games, and according to Ong (1982), also different cultures. The rules of the game differ; an eloquent speech may appear incoherent and repetitive in direct transcription, and an articulately argued article may sound boring when read aloud.

To transcribe means to transform, to change from one form to another. Attempts at verbatim interview transcriptions produce hybrids, artificial constructs that may be adequate to neither the lived oral conversation nor the formal style of written texts. Transcriptions are translations from an oral language to a written language; what is said in the hermeneutical tradition of translators also pertains to transcribers: *traduire traittori*—translators are traitors.

An interview is a live social interaction where the pace of the temporal unfolding, the tone of the voice, and the bodily expressions are immediately available to the participants in the face-to-face conversation, but they are not accessible to the out-of-context reader of the transcript. Especially irony is notoriously difficult to represent in a transcript, since it involves a deliberate discrepancy between nonverbal and verbal language, or between different elements of a verbal message—as Bourdieu has put it, irony as well as other common yet important tropes are almost certainly "lost in transcription" (1999, p. 622, note 15). The audio recording of the interview involves a first abstraction from the live physical presence of the conversing persons, with a loss of body language such as posture and gestures. The transcription of the interview conversation to a written form involves a second abstraction, where the tone of the voice, the intonations, and the breathing are lost. In short, transcripts are impoverished, decontextualized renderings of live interview conversations.

RECORDING INTERVIEWS

Methods of recording interviews for documentation and later analysis include audio recording, video recording, note taking, and remembering. The common way of recording interviews has been with the use of an audio recorder, which frees the interviewer to concentrate on the topic and the dynamics of the interview. The words and their tone, pauses, and the like are recorded in a permanent form that is possible to return to again and again for relistening. Today digital voice recorders are available; they provide a high acoustic quality and can record for many hours without interruption. The recordings can be transferred directly to a computer where they can be stored, played for analysis, and transcribed with the use of a word processor.

The first requirement for transcribing an interview is that it was in fact recorded. Some interviewers have painful memories of an exceptional interview where nothing got on the recording due to technical defects or, most often, human error. A second requirement for transcription is that the recorded conversation is audible to the transcriber. This may require that the interviewer takes measures to avoid background noise and is not afraid to ask a mumbling interviewee to speak up.

Video recordings offer a unique opportunity for analyzing the interpersonal interaction in an interview; the wealth of information, however, makes video analysis a time-consuming process. For ordinary interview projects, particularly those with many interviews and where the main interest is the content of what is said, video recordings may be too cumbersome for the analysis of the interview content. That said, video recordings of pilot interviews might be useful to sensitize interviewers to the importance of body language.

An interview may also be recorded by the interviewer's using remembrance, relying on his or her empathy and memory and then writing down the main aspects of the interview after the session, sometimes assisted by notes taken during the interview. Taking extensive notes during an interview may, however, be distracting, interrupting the free flow of conversation. There are obvious limitations to the interviewer's remembrance—such as a rapid forgetting of exact linguistic formulations, whereas the physical presence and the social atmosphere of the interview situation, lost on the recording, may remain in the background of memory. The interviewer's active listening and remembering may work as a selective filter, not only as a bias, but potentially also to retain those very meanings that are essential for the topic and the purpose of the interview. One might speculate that if audio recorders had been available at Freud's time, psychoanalytic theory might not have developed beyond infinite series of verbatim quotes from the patients, and psychoanalysis might today have remained confined to a small Viennese sect of psychoanalysts lost in the chaos of therapy tapes and disputes over their correct transcriptions.

TRANSCRIBING INTERVIEWS

Transcribing interviews from an oral to a written mode structures the interview conversations in a form amenable to closer analysis, and is in itself an initial analytic process. The amount and form of transcribing depends on such factors as the nature of the material and the purpose of the investigation, the time and money available, and—not to be forgotten—the availability of a reliable and patient typist.

Time and Resources for Transcription. The time needed to transcribe an interview will depend on the quality of the recording, the typing experience of the transcriber, and the demands for detail and exactitude. Transcribing large amounts of interview material is often a tiresome and stressing job, the stress being reduced by securing recordings of high acoustic quality. For the interviews in the grading study, an experienced secretary took about 5 hours to type a 1-hour interview. An interview of 1 hour results in 20 to 25 single-spaced pages, depending on the amount of speech and how it is set up in typing.

- *Who Should Transcribe?* In most interview studies, a secretary transcribes the recordings. Investigators who emphasize the modes of communication and linguistic style may choose to do their own transcribing in order to secure the many details relevant to their specific analysis. Mishler (1986; see also 1991) describes how he would let a secretary do a rough transcription of his interviews and then select a few interviews for extensive narrative analysis. He would transcribe these interviews himself, with linguistic annotations such as those in the craftsman narrative (Box 8.3). Researchers who transcribe their own interviews will learn much about their own interviewing style; to some extent they will have the social and emotional aspects of the interview situation present or reawakened during transcription, and will already have started the analysis of the meaning of what was said.

- *Transcription Procedure.* Transcribing from audio recording to text involves a series of technical and interpretational issues–in particular concerning verbatim oral versus written style—for which there are not many standard rules, but rather a series of choices to be made. There is one basic rule in transcription—state explicitly in the report how the transcriptions were made. This should preferably be based on written instructions to the transcribers. If there are several transcribers for the interviews of a single study, care should be taken that they use the same procedures for typing. If this is not done, it will be difficult to make linguistic cross-comparisons of the interviews.

• *Use of Transcriptions.* Although there is no universal form or code for transcription of research interviews, there are some standard choices to be made. Should the statements be transcribed verbatim and word-by-word, retaining frequent repetitions, noting "mh"s and the like, or should the interview be transformed into a more formal, written style? Sampling in interview studies does not only concern selection of subjects, transcription involves the sampling of which of the multiple dimensions of oral interview conversations are to be selected for written transcription; for example, should pauses, emphases in intonation, and emotional expressions like laughter and sighing be included? And if pauses are to be included, how much detail should be indicated? There are no correct, standard answers to such questions; the answers will depend on the intended use of the transcript, for example whether for a detailed linguistic or conversational analysis or for reporting the subject's accounts in a readable public story.

BOX 10.1 Transcription for Conversation Analysis

E: Oh honey that was a lovely luncheon I shoulda ca:lled you s:soo[:ner
   but I:]1:[lo:ved it.
M: [((f)) Oh:::] [(   )
E: It w's just deli:ghtfu[:1.   ]
M: [Well]
M: I w's gla[d   you] (came).]
E: ['nd yer f:] friends] 're so da:rli:ng,=
M: =Oh:::[:it w'z]
E: [e-that P]a:t isn't she a do:[:11?]
M: [iYe]h isn't she pretty.
                 (.)
E: <u>Oh</u>: she's a beautiful girl.=
M: =Yeh I think she's a pretty gir[1.=
E: [En' that Reinam'n::
                 (.)
E: <u>She</u> SCA:RES me.

SOURCE: J. Heritage, 1984, cited in *Doing Conversation Analysis* (p. 4), by P. ten Have,1999, Thousand Oaks, CA: Sage.

Transcription Conventions

[A single left bracket indicates the point of overlap onset.
]   A single right bracket indicates the point at which an utterance-
         part terminates vis-à-vis another.
                                                    (Continued)

(Continued)

=   Equal signs, one at the end of one line and one at the beginning of a next, indicate no "gap" between the two lines.

(.)   A dot in parentheses indicates a tiny "gap" within or between utterances.

::   Colons indicate prolongation of the immediately prior sound. Multiple colons indicate a more prolonged sound.

word   *Underscoring* indicates some form of stress, via pitch, and/or amplitude; an alternative method is to print the stressed part in *italics*.

WORD   *Upper case* indicates especially loud sounds relative to the surrounding talk.

( )   *Empty parentheses* indicate the transcriber's inability to hear what was said.

(( ))   *Double parentheses* indicate the transcriber's descriptions rather than transcriptions.

SOURCE: Adapted from *Doing Conversation Analysis* (Appendix), by P. ten Have, 1999, Thousand Oaks, CA: Sage.

The text in Box 10.1 may at first sight appear rather incomprehensible. It is a transcription of a sequence from a telephone conversation, presented here to demonstrate the complexities of transcription for special purposes, in this case for a conversation analysis, to be depicted in Chapter 13. We may also note that Mishler used some of the annotations in his rendering of a narrative interview with a craftsman (Box 8.3). Such specialized forms of transcription are neither feasible nor necessary for the meaning analysis of large interview texts in common interview projects. Whereas if the focus is on the linguistic style and the social interaction in a research interview, or in a doctor-patient interview, the pauses, overlaps, and intonations of the speech interaction may be of key importance. Transcriptions in detail, such as in Box 10.1, may also sensitize interviewers to the finer points of interview interaction.

The issue of how detailed a transcription should be is illustrated by the interview sequence below on competition for grades, which in Denmark is a negative behavior that many pupils hesitate to admit to.

Interviewer: Does it influence the relationship between the pupils that the grades are there?

Pupil: No, no-no, one does not look down on anyone who gets bad grades that is not done. I do not believe that: well, it may be that there are some who do it, but I don't.

Interviewer: Does that mean there is no competition in the class?

Pupil: That's right. There is none.

At face value, this boy says that one does not look down on pupils with low grades and confirms the interviewer's interpretation that there is no competition for grades in the class. A critical reading of the passage may, however, lead to the opposite conclusion—the repeated denials of looking down on other pupils may be interpreted as perhaps meaning the opposite of what was manifestly said. If the above interview sequence had not been transcribed verbatim, but rephrased into a briefer form such as "One does not look down on others with low grades nor compete for grades," the potential reinterpretation of the manifest meaning of the statement into its opposite could not have taken place.

In a study of the scholarly acculturation of university students of Danish and of medicine in their respective professional cultures, the interviews were transcribed verbatim, including the many "hm"s, "ain't it true," and the like (Jacobsen, 1981). The interviewer counted the use of such fillers by the students of Danish and of medicine, respectively, and found a markedly more frequent use of "ain't it true" by the students of Danish. He interpreted this, together with other indications, as being in line with the culture of the humanities, in which there is an emphasis on dialogue with attempts to obtain consensual validation of interpretations involving appeals to the others such as "ain't it true." In contrast, the medical profession is more characterized by lectures as monologues, authoritatively stating nondebatable truths.

TRANSCRIPTION RELIABILITY, VALIDITY, AND ETHICS

Transcriptions are constructions from an oral conversation to a written text. The constructive nature of transcripts appears when we take a closer look at their reliability and validity.

Reliability. Questions of interviewer reliability in interview research are frequently raised. Yet in contrast to sociolinguistic research, transcriber reliability is rarely mentioned in reference to social science interviews. Regarded technically, it is an easy check to have two persons independently type the same passage of a recorded interview, and then list and count the number of words that differ between the two transcriptions, thus providing a quantified reliability check.

The interpretational character of transcription is evident from the two transcripts of the same tape recording in Table 10.1. The words that are different in the two transcriptions are italicized. The transcriptions were made by two psychologists, who were instructed to transcribe as accurately as possible. Still, the transcribers adopted different styles: Transcriber A appears to write more verbatim, includes more words, and seems to guess more than transcriber B, who records only what is clear and distinct, and who also produces a more coherent written style. The most marked discrepancy between the two is rendering the interviewer's question as "because you don't get grades?" versus "*of course* you don't *like* grades?" It thereby becomes ambiguous what the subject's answer—"Yes, I think that's true . . ."—refers to.

Listening again to the recording might show that some discrepancies are due to poor recording quality, with mishearing and misinterpretations of hardly audible passages. Other divergences may not be unequivocally solved, as for example: Where does a sentence end? Where is there a pause? How long is a silence before it becomes a pause in a conversation? Does a specific pause belong to the subject or to the interviewer? And if the emotional aspects of the conversation are included, for instance "tense voice," "giggling," "nervous laughter," and so on, the intersubjective reliability of the transcription could develop into a research project of its own.

Table 10.1   Two Transcriptions of the Same Interview Passage

---

Transcription A

I: And are you also saying because you don't get grades? Is that true?

S: Yes, I think that's true because if I got grades I would work toward the grade as opposed to working toward . . . umm, expanding what I know, or, pushing a limit back in myself or, something . . . contributing new ideas . . .

Transcription B

I: And are you also saying *that of course* you don't *like* grades?

S: Yes, I think that's true, because if I got grades I would work toward the grade as opposed to working toward expanding what I know or pushing *those limits* back . . . (tape unclear) contributing new ideas.

Even the exact same written words in a transcript can convey two quite different meanings, depending on how the transcriber chooses to insert periods and commas. Poland (2003, p. 270) mentions the following example: "I hate it, you know. I do" (where "know" is not used to say anything about the hearer's knowledge), is very different from "I hate it. You know I do." (which assumes the hearer knows something). The very concept of a sentence fits with the tradition of written language and does not translate well into oral language, where we generally talk in flowing "run-on sentences" (p. 270). Where to insert periods and commas is already an interpretational process.

Validity. Ascertaining the validity of the interview transcripts is more intricate than assuring their reliability.

The issue of what a valid transcription is appears in the two different transcriptions of a story told by a 7-year-old African American pupil presented in Table 10.2. The two transcriptions are from a segment of a longer story from a classroom exercise, transcribed by two different researchers and discussed by Mishler (1991). Transcript A is a verbatim rendering of the oral form of the story; the teacher found the whole story disconnected and rambling, not living up to acceptable criteria of coherence and language use. Transcript B is an idealized realization of the same story passage, retranscribed into a poetic form by a researcher familiar with the linguistic practices of black oral style. Here the

Table 10.2   Two Transcriptions of Leona's Story of Her Puppy

---

Transcription A
. . . and then my puppy came / e was asleep / and he was—he was /
he tried to get up / and he ripped my pants / and he dropped the oatmeal—
all over him / and / my father came / and he said

. . .

Transcription B
an' then my puppy came
he was asleep
he tried to get up
an' he ripped my pants
an' he dropped the oatmeal all over him
an' my father came
an' he said

story appears as a literary tour de force, yielding a remarkable narrative. Neither transcription is more objective than the other; they are, rather, different written constructions from the same oral passage: "Different transcripts are constructions of different worlds, each designed to fit our particular theoretical assumptions and to allow us to explore their implications" (Mishler, 1991, p. 271).

Correspondingly, the question "What is the correct valid transcription?" cannot be answered—there is no true, objective transformation from the oral to the written mode. A more constructive question is: "What is a useful transcription for my research purposes?" Thus verbatim descriptions are necessary for linguistic analyses; the inclusion of pauses, repetitions, and tone of voice may also be relevant for psychological interpretations of, for example, level of anxiety or the meaning of denials. On the other hand, transforming the conversation into a literary style may highlight nuances of a statement and facilitate communication of the meaning of the subject's stories to readers.

The different rhetorical forms of oral and written language are frequently overlooked during the transcription of social science interviews; one exception is Poland (2003). Recognizing the socially constructed nature of the transcript, he discusses in detail procedures for increasing the trustworthiness of transcripts and thus enhancing rigor in qualitative research.

Neglecting linguistic complexities during transcription from an oral to a written language may be related to a philosophy of naive realism, with an implicit constancy hypothesis of some real meaning nuggets remaining constant by their transfer from one context to another. In contrast, postmodern conceptions of knowledge emphasize the contextuality of meaning, with an intrinsic relation of meaning and form, and focus on the very ruptures of communication, the breaks of meaning. The nuances and the differences, the transformations and discontinuities of meaning, become the very pores of knowledge. Postmodern approaches to knowledge do not solve the many technical and theoretical issues of transcription. The emphasis on the linguistic constitution of reality, on the contextuality of meaning, and on knowledge as arising from the transitions and breaks, however, involves sensitivity to and a focus on the transcription stage of interview research often overlooked.

Ethics. Transcription involves ethical issues. The interviews may treat sensitive topics in which it is important to protect the *confidentiality* of the subject and of persons and institutions mentioned in the interview. Among the necessary and simpler, but sometimes forgotten, tasks is the need for secure storage of recordings and transcripts, and of erasing the recordings when they are no longer of use. In sensitive cases, it may be advantageous as early as the transcription stage to mask the identities of the interviewed subjects, as well as events and persons in the interviews, that might be easily recognized.

Some subjects may experience a shock as a *consequence* of reading their own interviews. Oral language transcribed verbatim may appear as incoherent and confused speech, even as indicating a lower level of intellectual functioning. Some researchers may have painful memories of sending their transcripts back to the interviewees for corrections and comments, and receiving angry replies from offended subjects who may refuse further cooperation and any use of what they have said. If the transcripts are to be sent back to the interviewees, rendering them in a more fluent written style should be considered from the start. And if not, researchers should consider accompanying the transcripts with information about the natural differences between oral and written language styles. Be mindful that the *publication* of incoherent and repetitive verbatim interview transcripts may involve an unethical stigmatization of specific persons or groups of people.

Those teachers in the grading study who had expressed interest received a draft of the book chapter in which their statements were discussed. A teacher of Danish, who had been quoted extensively, called and asked the researcher (Steinar Kvale) to omit or rephrase his statements in the book. The rather off-the-cuff verbatim and highly illustrative quotes from his interview showed very poor Danish language used by a teacher of Danish, which he found embarrassing in his profession. At that time the researcher was little aware of the different rules for oral and written language and believed that a verbatim transcription of the interviews was the most loyal and objective transcription. However, the teacher's request was respected and his quotes were changed into a correct written form, which also made them more readable.

What is verbatim?

ELEVEN

PREPARING FOR
INTERVIEW ANALYSIS

—◆—

We will address how to analyze interviews in this and the three follow-ing chapters, focusing on, respectively, the analysis of meaning (Chapter 12), linguistic analyses (Chapter 13), and theoretical analyses (Chapter 14). We introduce the analysis of interviews in this chapter by addressing a question, which may be posed by novice researchers who are about to begin analyzing their interview material. After that we present steps and modes of analysis and then briefly discuss the use of computers and soft-ware in analyzing qualitative interviews.

THE 1,000-PAGE QUESTION

In many cases, the practical problems of an interview project reach their high-point when the researcher is confronted with his or her many pages of inter-view transcripts and is led to ask the question:

How shall I find a method to analyze the 1,000 pages of interview tran-scripts I have collected?

Before addressing in greater detail how to analyze interviews in the fol-lowing chapters, we will attempt to give a reply to this 1,000-page question. The answer includes some summaries of the stages of an interview investigation that have already been covered, and prepares the ground for the analysis stage itself.

A first impulsive reaction to the 1,000-page question above is to dismiss it—"Never pose that question!" When an interview project has been conducted in such a way that the 1,000-page question is asked, the question can no longer be answered. A more adequate reply would then be: "Never conduct interview research in such a way that you find yourself in a situation where you ask such a question."

In what follows, we will use a brief analysis of the 17 words in the 1,000-page question formulated above to illustrate concretely how one can analyze textual material. The purpose of the analysis will be to uncover the meaning of the question, to make explicit its presuppositions and thereby the implicit conceptions of qualitative research it implies. The general purpose is prophylactic; it is an attempt to outline modes of conducting interview research so that a researcher never gets into a situation where he or she feels compelled to ask the 1,000-page question. The general form of the analysis is to select 7 key words from the 1,000-page question and analyze them separately:

How shall I find a *method to analyze* the *1,000 pages* of interview *transcripts* I *have collected*?

"*Have*"—*too late!* The question is posed too late: Never pose the question of how to analyze transcripts *after* the interviews have been conducted—it is too late to start thinking after the interviewing is done. The answer here parallels that of a statistician: Consult me about the data analysis before you collect your data.

Think about how the interviews are to be analyzed before they are conducted. The method of analysis decided on—or at least considered—will then guide the preparation of the interview guide, the interview process, and the transcription of the interviews. In addition, the analysis may also, to varying degrees, be built into the interview situation itself. In such forms of analysis—interpreting "as you go"—considerable parts of the analysis are "pushed forward" into the interview situation itself. The final analysis then becomes not only easier and more amenable, but also rests on more secure ground. Put strongly, the ideal interview is already analyzed by the time the sound recorder is turned off. There are social and ethical restraints on how far the analysis of meaning can be undertaken during the interview itself, but this may serve as a methodological ideal for interview research. Socrates' conversations with the citizens of Athens that have been addressed in previous chapters represent an extreme form of immanent analysis-in-situ, resulting in self-communicating interviews in a readable literary style.

A reformulation of the 1,000-page question entails changing the temporal form: *How shall I conduct my interviews so that their meaning can be analyzed in a coherent and creative way?*

"1,000 pages"—too much! The answer to this quantitative part of the question is simple—1,000 pages of transcripts are more often than not too much to handle in a meaningful way.

The precise meaning of the question may depend on its intonation. When posed in a despairing voice, it may indicate a situation of being overwhelmed by an enormous amount of qualitative data, of being completely lost in a jungle of transcriptions. The meaning of the question may then be: "Rescue me from my 1,000 pages, I cannot find my way out of the labyrinth." When posed in a more assertive voice, the same question may have another meaning. A diligent young scholar has done his empirical duty and documented his scientific attitude by gathering large amounts of data. He now awaits the expert's praise and advice about how to treat the data. The question may here involve a "qualitative positivism"—a quest for scientific respectability by mirroring the positivist emphasis on large quantities of quantitative data with large quantities of qualitative data. Whether posed in a despairing or in an assertive voice, the formulation of the question leads in the wrong direction. The emphasis is on the quantity—1,000 pages—rather than on the content and the qualitative meanings of what was said.

A rephrasing of the 1,000-page question, involving a change in emphasis from quantity to meaning, could be: *How do I go about finding the meaning of the many interesting and complex stories my interviewees told me?*

"How"—ask "what" and "why" first. Do not pose the question of "how" to analyze interviews before the answers to the "what" and the "why" of an investigation have been given. Content and purpose precede method.

Method originally meant the way to the goal. With no goal stated, it is difficult to show the way to it. In general, theoretical conceptions of what is investigated should provide the basis for making decisions of how—the method to be used for analyzing the content. Further, if a research study purports to test a hypothesis about differences among groups of subjects, then the analysis should be systematic and conducted in the same way for each of the groups in order to test possible differences among them. For explorative purposes it will, on the contrary, be more appropriate to pursue the different interesting aspects of the individual interviews and to interpret them in greater depth.

The technical "how to" emphasis of the 1,000-page question can be reformulated to read: *How do I go about finding out what the interviews tell me about what I want to know?*

"Method" versus knowledge. The methodological aspect of the 1,000-page question cannot be answered due to the way the question is formulated. As we argued in Chapter 5, there are no standard methods, no *via regia,* to arrive at the meaning of what is said in an interview. Rather, such understanding is based on the experience and the craftsmanship of the researcher. The search for techniques of analysis may be a quest for a "technological fix" to the researcher's task of analyzing and constructing meaning. No standard methods of text analysis exist that correspond to the multitude of techniques available for statistical analysis. The alternative to a methodical emphasis of the 1,000-page question is: *How can the interviews assist me in extending my knowledge of the phenomena I am investigating?*

"Transcripts"—beware! Do not conceive of the interviews as transcripts—the interviews are living conversations.

The transcripts should not be the subject matter of an interview study, as implied by the 1,000-page question, but rather be means or tools for the interpretation of what was said during the interviews. The transcript is a bastard, a hybrid between an oral discourse unfolding over time, face to face, in a lived situation—where what is said is addressed to a specific listener present—and a written text created for a general, distant public. An emphasis on the transcription may promote a reifying analysis that reduces the text to a mere collection of words or single meanings conceived as verbal data. An alternative approach to the transcripts involves entering into a *dialogue* with the text, going into an imagined conversation with the "author" about the meaning of the text. The reader here asks about the theme of the text, goes into the text seeking to develop, clarify, and expand what is expressed in the text. The alternative to the transcription emphasis in the 1,000-page question is: *How do I analyze what my interviewees told me in order to enrich and deepen the meaning of what they said?*

"Collected" versus co-authored. The interviewee's statements are not collected—they are co-authored by the interviewer.

The interview is an inter-subjective enterprise of two persons talking about common themes of interest. The interviewer does not merely collect statements like gathering small stones on a beach. His or her questions lead up to what aspects of a topic the subject will address, and the interviewer's active listening and following up on the answers co-determines the course of the conversation. There is a tendency to take the results of a social interaction, when first arrived at, as something given, forgetting the original discourse and the social co-construction of the final outcomes. Such reification may be strengthened by the transcription of the interviews; the fixed written form takes over and the original face-to-face interaction of the interview situation fades away.

The role of the interviewer as a co-producer and a co-author of the interview, and in reflecting on the social constitution of the interview, is then overlooked. Focusing on the transcripts as a collection of statements may freeze the interview into a finished entity rather than treat its passages as stepping-stones toward a continuous unfolding of the meaning of what was said. The analysis of the transcribed interviews is a continuation of the conversation that started in the interview situation, unfolding its horizon of possible meanings.

The alternative to the stamp-collecting version of the 1,000-page question is: *How do I carry on the dialogue with the text I have co-authored with the interviewee?*

"Analyze" versus narrate. Do not let the analysis stage inflate so that it consumes the major portion of time available for an interview project.

The analysis of an interview is interspersed between the initial story told by the interviewee to the researcher and the final story told by the researcher to an audience. To analyze means to separate something into parts or elements. The transcription of the conversation and the conception of the interview as a collection of statements might suggest a fragmentation of the story told by the interviewee into separate parts, be they single paragraphs, sentences, or words. It is then easy to forget that in open, non-directive interviews the interviewee tells a story, or several stories, to the researcher, and that the transcript itself may then approximate the form of a narrative text.

A narrative approach to the interview analysis, going back to the original story told by the interviewee and anticipating the final story to be reported to an audience, may prevent becoming lost in a jungle of transcripts. A focus on the interview as a narrative may even make the interview transcripts better reading, given that the original interview is deliberately created in a story form. A narrative conception of interview research supports a unity of form among the original interview situation, the analysis, and the final report.

A narrative alternative to the analysis version of the 1,000-page question then becomes: *How can I reconstruct the original story told to me by the interviewee into a story I want to tell my audience?*

Method of Analysis

Our answer to the 1,000-page question has been that the question was posed too late to obtain a satisfactory answer and that its formulation made it difficult to answer directly. The wording of the question was then analyzed in detail with the purpose of bringing its implicit presuppositions of interview research into the open, and with the general interest of making the question superfluous.

No standardized method of analysis was applied to the question; rather, a variety of approaches were tried in order to bring out the meaning of the question. The general structure was to select 7 key words from the 17-word sentence and examine them separately. Yet the analysis was not entirely decontextualized; there was continuous overlapping among the meanings developed from the key words that pointed to common threads of meaning underlying the question. When probing the meaning of the separate words, an attempt was made to bring in the context of the question.

The original sentence was rephrased in various forms, leading to different directions of meanings. The alternative rephrasings of the 1,000-page question shifted the focus from what was said to what could have been said, opening up some of the possibilities of meanings that the original formulation of the question closed off.

The deconstruction of the 1,000-page question involved a destruction of the presuppositions of the question and a construction of alternative formulations for enriching interview analysis. The interpretation addressed the tension between what was said and what was not said in the question. This interplay of the said and the not-said did not lead to one, true, objective meaning of the question, but served to keep the conversation going about the meanings of interview analysis the question opens up.

The topic of the original question concerned 1,000 pages of interview transcripts of questions and answers, and it was postulated that this was too much material for undertaking a comprehensive analysis. The above interpretation of the 17 words of the 1,000-page question has required around 2,134 words, which makes the quantitative relation of original text to interpretative text 1:126. A corresponding interpretation of the meaning of 1,000 pages would then require 126,000 pages, which amount to somewhere between 500 and 1,000 books.

STEPS AND MODES OF INTERVIEW ANALYSIS

In the following three chapters, we present different modes of interview analysis, which will possibly disappoint those who expect magic wands that finally uncover the treasures of meaning hidden in their many pages of transcripts. Some common approaches to the analysis of the meaning of interview texts—involving different technical procedures—do exist, though. The techniques of analysis are tools that are useful for some purposes, relevant for some types of interviews, and suited for some researchers. Before turning to the specific tools, we will outline some more or less common steps in interview analyses.

BOX 11.1   Six Steps of Analysis

A first step is when *subjects describe* their life world during the interview. They spontaneously tell what they experience, feel, and do in relation to a topic. There is little interpretation or explanation from either the interviewees or the interviewer.

A second step would be that the *subjects themselves discover* new relationships during the interview, see new meanings in what they experience and do on the basis of their spontaneous descriptions, free of interpretation by the interviewer. For example, a pupil, describing the effects of grading, comes to think of how the grades further a destructive competition among pupils.

In a third step, the *interviewer, during the interview, condenses and interprets* the meaning of what the interviewee describes, and "sends" the meaning back. The interviewee then has the opportunity to reply, for example, "I did not mean that" or "That was precisely what I was trying to say" or "No, that was not quite what I felt. It was more like. . . ." This process ideally continues till there is only one possible interpretation left, or it is established that the subject has multiple, and possibly contradictory, understandings of a theme. This form of interviewing implies an ongoing "on-the-line interpretation" with the possibility of an "on-the-spot" confirmation or disconfirmation of the interviewer's interpretations. The final product can then be a "self-correcting" interview.

In a fourth step, the *recorded interview is analyzed by the interviewer* alone, or with co-researchers. The interview is usually structured for analysis

(Continued)

---

lack of = lack of reliability

(Continued)

by transcription and with the aid of computer programs for textual analysis. The analysis proper involves developing the meanings of the interviews, bringing the subjects' own understanding into the light as well as providing new perspectives from the researcher. A variety of analytical tools focusing on the meaning and the linguistic form of the texts is available.

A fifth step could be a *reinterview*. When the researcher has analyzed the interview texts, he or she may give the interpretations back to the subjects. In a continuation of a "self-correcting" interview, the subjects then get an opportunity to comment on the interviewer's interpretations as well as to elaborate on their own original statements, as a form of "member validation."

A possible sixth step would be to extend the continuum of description and interpretation to include *action*, by subjects beginning to act on new insights they have gained during their interview. In such cases the research interview may approach the form of a therapeutic interview. The changes can also be brought about by collective actions in a larger social setting such as action research, where researcher and subjects together act on the basis of the knowledge they have produced in the interviews.

Box 11.1 depicts six steps of a continuum from description to interpretation and action, which do not necessarily presuppose each other. The first three steps of description, discovery, and interpretation in the interview situation were outlined in Chapters 7 and 8. The following overview and descriptions of main analytic tools can assist interviewers in choosing modes of analysis adequate for their project. The quality of the analysis rests upon the craftsmanship of the researcher, his or her knowledge of the research topic, sensitivity to the medium he or she is working with—language, and mastery of tools for analysis. Some of the analytic tools, such as coding, have been developed through practice without a theoretical basis; some have been inspired by philosophical traditions, such as phenomenology and hermeneutics; and some, such as discursive and deconstructive analyses, have been derived directly from specific philosophical positions.

Some key approaches to the analysis of interview texts are presented in Table 11.1. Meaning and language are intertwined. Still, for practical purposes, we divide the modes of analysis into a first group of analyses, which mainly focus on the meaning of what is said, treated in Chapter 12, and a second group of analyses, which mainly work with the linguistic forms whereby

Table 11.1   Modes of Interview Analysis

Analyses Focusing on Meaning
Meaning coding
Meaning condensation
Meaning interpretation
Analyses Focusing on Language
Linguistic analysis
Conversation analysis
Narrative analysis
Discourse analysis
Deconstruction
General Analyses
Bricolage
Theoretical reading

meanings are expressed, in Chapter 13. For the approaches in Chapter 12, we go into detail by outlining how to proceed in practice, whereas for the wide variety of analytic tools presented in Chapter 13, we mainly provide some examples and refer to further literature for their application. Finally, we turn in Chapter 14 to more general approaches to analysis, such as bricolage, an eclectic combination of multiple forms of analysis, and a theoretically informed reading of the interviews as a significant mode of analysis. With the widely different techniques of analysis, and their frequently diverging epistemological assumptions, the following three chapters may make for challenging reading. Their purpose, though, is to give a brief overview of some key tools for interview analysis, which the interview researcher may choose among according to the content and purpose of his or her inquiry.

Some of the analyses come close to a traditional understanding of knowledge as preexisting elements that can be collected, such as meaning coding and condensation, attempting to bring out what is already there in the texts. Other analyses are more in line with a view of knowledge as socially constructed, such as discursive and deconstructive analyses, which focus on the language medium of the stories told, generally without taking a position on whether the conversations refer to any objective data or essential meanings.

For the varieties of meaning interpretation and narrative analysis, both the miner and traveler metaphors of interview knowledge may apply.

COMPUTER TOOLS FOR INTERVIEW ANALYSIS

Among novices to the trade of interview research there may be a belief that the intricate analysis of their interviews will be taken care of by computer software. The marketing of computer programs for textual analysis feeds into such high expectations, for example:

> Qualitative research is a challenge!
>
> You are faced with rich data from interviews, focus groups, observations, surveys, profiles or web-searches. How do you make sense of it and do justice to it? How do you find and explain patterns, identify themes and powerfully deliver your results?
>
> Meet the challenge!
>
> XXX offers the world's leading qualitative solutions for researchers.

Program developers themselves and textbook authors may promise less about computer software for textual analysis: "They are not a substitute for thought, but they are a strong aid for thought . . . Computers don't analyse data; people do" (Weitzman & Miles, 1995, p. 3). Computer programs can facilitate the analysis of interview transcripts. They replace the time-consuming "cut-and-paste" approach to hundreds of pages of transcripts with "electronic scissors." The programs are aids for structuring the interview material for further analysis; the task and the responsibility of interpretation, however, remain with the researcher. The computer programs allow for such operations as writing memos, writing reflections on the interviews for later analyses, coding, searching for key words, doing word counts, and making graphic displays. Some of the programs allow for on-screen coding and taking notes while reading the transcripts. It should be noted that several of these tasks may also be taken care of by common word processors.

The most common form of computer analysis today is coding, or categorization, of the interview statements, to be discussed in Chapter 12. When coding, the researcher first reads through the transcripts and codes the relevant passages; then, with the aid of code-and-retrieve programs, the coded passages can be retrieved and inspected over again, with options of recoding and of combining codes. Existing computer programs have been developed for coding strategies (see, e.g., the overview of different software packages in Gibbs, 2007), whereas computer-assisted programs for textual analysis have not been developed for many other forms of interview analysis, such as narrative and discursive analyses. There is thus a danger that the ready availability of computer programs for coding can have the effect that coding becomes a preferred shortcut to analysis, at the expense of a rich variety of modes of interview analyses, to be presented in the following chapters.

While most computer programs today work with written texts, newer forms exist that deal directly with the sound recording. The investigator can listen directly to the interview interaction, code it, and write associations or interpretations on the text screen and later immediately get back to the coded sound sequence. Working directly with the sound will save time and money for transcribing entire interviews, as well as overstep many of the problems of transcription discussed above and secure the researcher a close contact with the original oral discourse. Smaller passages to be reported or analyzed more intensively may then be selected for transcription. Either way, as Poland (2003, p. 284) reminds us: "It is unlikely that technology will enable researchers to bypass the thorny issues of interpretation involved in the preparation of data for analysis." And he adds the question: "Would we really want it any other way?"

TWELVE

INTERVIEW ANALYSES FOCUSING ON MEANING

———⋅◆⋅———

Tools exist that can make the interview analysis more accessible than it
seems from the reply to the 1,000-page question. In this and the next two
chapters, we give an overview of analytic tools and approaches for analyzing
interview texts—with a key purpose of sensitizing interviewers when they are
conducting and transcribing their interviews to the specific demands that different modes of analysis pose to interviews and transcriptions.

In this chapter, we depict in some detail the application of procedures for
coding, condensation, and—in particular—interpretation of meaning. They
can be used to organize the interview texts, to concentrate the meanings into
forms that can be presented in a relatively short space, and to work out implicit
meanings of what was said. Coding and condensation provide structure and
give overviews to often extensive interview texts. Meaning interpretation may
focus on small segments of interaction—such as from a therapy session and
from an interview on grading—and through multiple interpretations extend the
original texts. Inspired by hermeneutic text interpretation, we emphasize the
primacy of the question: how the interpreters' presuppositions and questions
put to the text co-constitute the meanings interpreted.

MEANING CODING

Coding and categorizing were early approaches to the analysis of texts in the
social sciences. Coding involves attaching one or more keywords to a text

segment in order to permit later identification of a statement, whereas categorization entails a more systematic conceptualization of a statement, opening it for quantification. However, the two terms are often used interchangeably. In various forms, coding is a key aspect of grounded theory, content analysis, and computer-assisted analyses of interview texts (see Gibbs, 2007).

Coding played an important role in the *grounded theory* approach to qualitative research introduced by Glaser and Strauss in 1967. Here *open coding* refers to "the process of breaking down, examining, comparing, conceptualizing and categorizing data" (Strauss & Corbin, 1990, p. 61). In contrast to content analysis, the codes in a grounded theory approach do not need to be quantified, but enter into a qualitative analysis of the relations to other codes and to context and action consequences. Coding has also become a key feature of the new programs for computer-assisted analysis of interviews (see Chapter 11).

The original formulations of Glaser and Strauss were significant in enabling qualitative researchers to provide legitimation for their methods of inquiry by presenting explicit strategies in the form of a codified research process. The purpose of grounded theory is not to test existing theory, but to develop theory inductively. Concretely, working with grounded theory implies a thorough coding of the material, and, as presented by Charmaz (2005), codes are immediate, are short, and define the action or experience described by the interviewee. The goal is the development of categories that capture the fullness of the experiences and actions studied. Data instances are constantly compared for similarities and differences, which lead to sampling of new data and writing of theoretical memoranda. More focused coding is then undertaken, and the analysis is gradually moved from descriptive to more theoretical levels, leading to a "saturation" of the material by the coding process, when no new insights and interpretations seem to emerge from further codings.

As Gibbs (2007) makes clear, coding demands the use of "code memos" where the researcher records the names of the different codes, who coded which parts of the material, the date when the coding was done, definitions of the codes used, and notes about the researcher's thoughts about the code (p. 41). Coding can be either concept driven or data driven. Concept-driven coding uses codes that have been developed in advance by the researcher, either by looking at some of the material or by consulting existing literature in the field, whereas data-driven coding implies that the researcher starts out without codes, and develops them through readings of the material (e.g., in grounded theory). In principle, anything can be coded—Gibbs suggests the following examples: specific acts, events, activities, strategies, states, meanings, norms, symbols, levels of participation, relationships, conditions or constraints, consequences, settings, and also reflexive codings, recording the researcher's role in the process (pp. 47–48).

Content analysis is a technique for a systematic quantitative description of the manifest content of communication. It was developed for the study of enemy propaganda during World War II and has since been used extensively for media analysis. The coding of a text's meaning into categories makes it possible to quantify how often specific themes are addressed in a text, and the frequency of themes can then be compared and correlated with other measures.

When coding takes the form of *categorization,* the meaning of long interview statements is reduced to a few simple categories. When coding into fixed well-defined categories, the occurrence or nonoccurrence of a phenomenon can be expressed by a simple "+" or "–" symbol. The strength of an opinion can also be indicated with a single number on a scale of, for example, 1 to 7. Categorization thus reduces and structures large interview texts into a few tables and figures. The categories can be developed in advance or they can arise ad hoc during the analysis; they may be taken from theory or from the vernacular, as well as from the interviewees' own idioms. Categorizing the interviews of an investigation can provide an overview of large amounts of transcripts, and facilitate comparisons and hypothesis testing.

In the interview study by Kinsey and his colleagues (1948; see Box 6.6), the interviewees' statements were not recorded in longhand, but coded during the interview with respect to several hundred themes, which had been defined in advance of the interviews. There are thus no qualitative renderings of the contents of the interviews; all of the findings were reported quantitatively in a multitude of tables. Coding during the interviews made it possible to record as rapidly as one can carry out a conversation, without loss of rapport or blockage on the subjects' parts, which made it possible to secure a complete history without slowing down an interview. Recording during the interview increased the accuracy of the coding, because the subject was present at the time and additional clarifying questions could be asked on the spot. The coding in situ further contributed to securing the anonymity of the subjects. While there have later been critiques of the statistical findings as based on nonrandom selections of subjects, the use of qualitative interviews in the study has received little attention.

The analysis of the interviews on grades illustrates one form of categorization based on transcribed interviews. The 30 pupil interviews, transcribed into 762 pages, were categorized in order to test the hypothesis that using grades to measure learning affects learning and social relations in school. Figure 12.1 depicts eight subcategories of one main dimension of a grade perspective—"Relationship with the teacher" (the other main dimensions were: Relation to fellow pupils, Self-concept, Relations to time, Emotional relations, Learning motivation, and Learning form). The categories were taken from educational literature and pilot interviews, and defined (for example: "Bluffing—the pupil attempts to give the impression that he knows more than he knows, with the purpose of obtaining better grades" and "Wheedling—the pupil attempts to win the sympathy of the teacher with the purpose of obtaining better grades"). Two coders independently categorized the 30 interviews, and their codings were combined. Figure 12.1 depicts how many of the 30 pupils confirmed or disconfirmed the occurrence of each of the eight subcategories of the dimension "Relationship with the teacher," generally supporting the hypothesis that grades influence social relations in school (Kvale, 1980).

The categorization of the meanings of the pupils' statements serves several purposes: (a) The categorizations structured the extensive and complex interviews and when presented in a figure gave a simple overview of the occurrence of grading behaviors among the 30 pupils interviewed. Thus in seven figures, as shown in Figure 12.1, the main results of 762 pages of interview

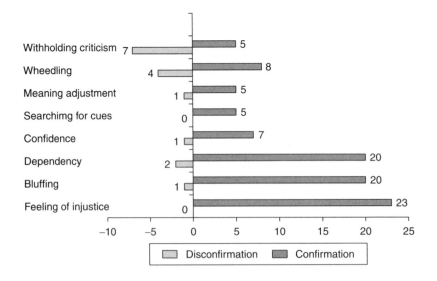

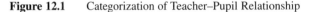

Figure 12.1 Categorization of Teacher–Pupil Relationship

transcription regarding the extent of grading attitudes and behaviors could be reported. (b) The categorization made it possible to test the hypothesis that grades influence learning. (c) The overviews of categorizations, such as those in Figure 12.1, give readers a background for judging how typical the quotes used in the accompanying qualitative analyses were for the interview material as a whole, and the categorizations may to a certain extent serve to counterbalance selective interpretations. (d) The categorization made it possible to investigate differences in grading behaviors for different groups among the 30 pupils, such as boys versus girls and pupils with high versus low grades. (e) Quantification also made comparisons to other investigations on the effects of grades possible. (f) The categorization could itself be checked for coder reliability, which made some checks for interviewer reliability possible.

Implictions for Interviewing: With categorization involving either/or decisions, it is preferable with precise preinterview definitions of the categories and careful probing during the interview to ascertain how the statements may be categorized. When the codes or categories are not to be developed until interviewing and analysis, it is important during the interviews to obtain rich descriptions of the specific phenomena to be coded or categorized.

MEANING CONDENSATION

Meaning condensation entails an abridgement of the meanings expressed by the interviewees into shorter formulations. Long statements are compressed into briefer statements in which the main sense of what is said is rephrased in a few words. We shall here exemplify one form of meaning condensation developed by Giorgi (1975) on the basis of phenomenological philosophy. The thematic purpose of his study was to investigate what constitutes learning for ordinary people in their everyday activities. The methodological purpose was to demonstrate how one deals systematically with data that remain expressed in terms of ordinary language and how rigor and discipline can be applied in data analysis without necessarily transforming the data into quantitative expressions.

Table 12.1 demonstrates how the interview about learning presented in Chapter 2 was subjected to meaning condensation. The analysis of an interview involves five steps. First, the complete interview is read through to get a sense of the whole. Then, the natural "meaning units" of the text, as they are expressed by the subjects, are determined by the researcher. Third, the theme

Table 12.1	Meaning Condensation

Natural Unit	Central Theme
1. The first thing that comes to mind is what I learned about interior decorating from Myrtis. She was telling me about the way you see things. Her view of looking at different rooms has been altered. She told me that when you come into a room you don't usually notice how many vertical and horizontal lines there are, at least consciously, you don't notice. And yet, if you were to take someone who knows what's going on in the field of interior decoration, they would intuitively feel there was the right number of vertical and horizontal lines.	1. Role of vertical and horizontal lines in interior decorating
2. So, I went home, and I started looking at the lines in our living room, and I counted the number of horizontal and vertical lines, many of which I had never realized were lines before. A beam . . . I had never really thought of that as vertical before, just as a protrusion from the wall. (Laughs)	2. S looks for vertical and horizontal lines in her home
3. I found out what was wrong with our living room design: many, too many, horizontal lines and not enough vertical. So I started trying to move things around and change the way it looked. I did this by moving several pieces of furniture and taking out several knick-knacks, de-emphasizing certain lines, and . . . it really looked differently to me.	3. S found too many horizontal lines in living room and succeeded in changing its appearance
4. It's interesting because my husband came home several hours later and I said "Look at the living room, it's all different." Not knowing this, that I had picked up, he didn't look at it in the same way I did. He saw things were moved, but he wasn't able to verbalize that there was a de-emphasis on the horizontal lines and more of an emphasis on the vertical. So I felt I learned something.	4. Husband confirms difference, not knowing why

SOURCE: "An Application of Phenomenological Method in Psychology," in A. Giorgi, C. Fischer, & E. Murray (Eds.) (1975), *Duquesne Studies in Phenomenological Psychology* (Vol. 2, pp. 82–103), Pittsburgh, PA: Duquesne University Press.

that dominates a natural meaning unit is restated by the researcher as simply as possible, thematizing the statements from the subject's viewpoint as understood by the researcher. Table 12.1 depicts this third step of analysis. The fourth step consists of interrogating the meaning units in terms of the specific purpose of the study. In the fifth step, the essential, nonredundant themes of the entire interview are tied together into a descriptive statement.

This form of meaning condensation can serve to analyze extensive and often complex interview texts by looking for natural meaning units and explicating their main themes. These themes may thereafter be subject to more extensive interpretations and theoretical analyses. Giorgi thus points out the importance of interpersonal relations in learning, which emerged in this study, a phenomenon that was rather neglected in the theories of learning at the time. It should also be noted that meaning condensation is not confined to a phenomenological approach and is also applied in other qualitative studies (Tesch, 1990).

Implications for Interviewing: For a phenomenologically based meaning condensation, it becomes paramount to obtain rich and nuanced descriptions of the phenomena investigated in the subjects' everyday language. The interviewer's theories of the subject matter should be "put into brackets" during the interview, in line with a phenomenological approach outlined in Chapter 2.

MEANING INTERPRETATION

The interpretation of the meaning of interview texts goes beyond a structuring of the manifest meanings of what is said to deeper and more critical interpretations of the text. Meaning interpretation is prevalent in the humanities, such as in a critic's interpretations of a poem or a film, and in psychoanalytical interpretations of patients' dreams. The interpreter goes beyond what is directly said to work out structures and relations of meanings not immediately apparent in a text. In contrast to the decontextualization of statements through categorization, interpretation recontextualizes the statements within broader frames of reference. As compared to the text reduction techniques of categorization and condensation, interpretations often lead to a text expansion, with the outcome formulated in far more words than the original statements interpreted (e.g., in the interpretations of Hamlet's interview in Chapter 9 and the 1,000-page question in Chapter 11).

The interpretation of the meaning of texts encompasses a variety of approaches. Here we first exemplify multiple interpretations in a therapeutic example, then present a hermeneutic approach to the meaning of texts, and finally discuss meaning interpretation in relation to the questions a researcher poses to the interview texts.

THE ISSUE OF MULTIPLE INTERPRETATIONS

The evaluation of the quality of Hamlet's interview showed how different readings of an interview resulted in rather different interpretations, such as whether Hamlet's leading questions lead to unreliable or reliable knowledge, and whether it is Polonius or Hamlet who is fooled in the interview. No systematic method of meaning interpretation was in play here. The interpretation of the 1,000-page question unfolded multiple potential meanings of the question, and here the methodical approach was spelled out. We will now address the issue of multiple interpretations with an example from a family therapy (Scheflen, 1978).

BOX 12.1   Susan's Smile

The issue of multiple interpretations is brought out by a case story on family interaction in a therapy. It is cast in a narrative form with a group of therapists watching and commenting on a videotaped therapy session. At one point the daughter, Susan, had smiled in an enigmatic way. The discussion among the observers about the meaning of this nonverbal statement, leading to six different interpretations, can also highlight issues of interview interpretations.

   One therapist suggested that the smile was sarcastic, thus invoking an expressional paradigm, where a person's actions are attributed to something within the person. Then a member of the group offered a second interpretation by pointing out that just before Susan had smiled her father had turned to her, held out his hands, and said "I think Susan loves us. We certainly love her." The smile was then seen as a response to her father's statement. A further observation led to a third interpretation: After Susan had smiled, her mother turned to her and said: "You never appreciate what we try to do for you." The smile was then interpreted as a provocation, as a stimulus for the mother's reprimand. In these three explanations Susan's smile was interpreted as an expression, as a response, and as a stimulus, focusing respectively on Susan in isolation, the father-daughter relationship, and the mother-daughter relationship.

A fourth interpretation followed from a closer focus on the interpersonal interaction, noticing that the three members of the family often acted and reacted to each other by withdrawal: When Susan smiled her father turned his face away and fell silent, and when the mother began her reprimand Susan reacted in a similar way. A fifth interpretation followed when the tape was played back and the therapists looked for incidents similar to the sequence in which Susan smiled. There had been two previous exchanges where the father approached, Susan smiled, and the mother reprimanded. This indicated a programmed interaction in this family, the actors following an unwritten script and interacting according to a preexisting scenario. In this interpretation, moving from an individual-centered to a cultural interpretation, Susan smiled because this was the part she was expected to play in the family drama. A sixth interpretation argued that although Susan's smile was a response to her father's approach, it was not a response in kind. In Bateson's language, the smile was meta to the father's statement, her metacommunication derailed her father's offer of involvement.

When discussing the six therapists' interpretations of Susan's smile, Scheflen (1978, p. 59) does not side with any one interpretation: "These are usually presented as opposing truths in different doctrinal schools, but they are all valid from one point of view or another. And, accordingly, they are all tactically useful at some point or another." The various modes of explanation can be used deliberately as tactics throughout a therapy, can be tactically employed to alter habitual tendencies to deny, ignore, project, and blame: "In the course of family therapy our clients can learn multiple approaches from us and end up with a more flexible and comprehensive strategy for viewing and making sense of their experiences" (p. 68).

The case of Susan, who smiles enigmatically in the course of a therapy session, is outlined in Box 12.1. The therapy sequence testifies to the vague, uncertain, and precarious nature of meanings in interpersonal interactions, which may also pertain to the meanings produced in qualitative research interviews. The therapists discussing the meaning of Susan's smile offer continually new interpretations as new contexts—from the therapy and from theory—are drawn in. The different interpretations do not necessarily contradict each other; they may be seen as enriching the meaning of a vague behavior. In the therapeutic session they may all be valid in one form or another, and they may be tactically useful to the therapeutic task of assisting the family in changing their habitual forms of interaction.

HERMENEUTICAL INTERPRETATION OF MEANING

We now turn to the hermeneutical tradition in the humanities, which for centuries has sought to come to grips with the vicissitudes of the interpretations of texts, notably the Bible, and also legal and literary texts. We begin by outlining some hermeneutic canons of interpretations.

BOX 12.2   Hermeneutical Canons of Interpretation

The first canon involves the continuous back-and-forth process between parts and the whole, which follows from the *hermeneutical circle*. Starting with an often vague and intuitive understanding of the text as a whole, its different parts are interpreted, and out of these interpretations the parts are again related to the totality, and so on. In the hermeneutical tradition this circularity is not viewed as a "vicious circle," but rather as a "circulus fructuosis," or spiral, which implies a possibility of a continuously deepened understanding of meaning. The problem is not to get away from the circularity in the explication of meanings, but to get into the circle in the right way.

A second canon is that an interpretation of meaning ends when one has reached a "good Gestalt," an inner unity of the text, which is free of logical contradictions.

A third canon is the *testing* of part-interpretations against the global meaning of the text and possibly also against other texts by the same author.

A fourth canon is the *autonomy of the text;* the text should be understood on the basis of its own frame of reference by explicating what the text itself states about a theme.

A fifth canon of the hermeneutical explication of a text concerns *knowledge about the theme* of the text.

A sixth principle is that an interpretation of a text is not *presupposition-less*. The interpreter cannot "jump outside" the tradition of understanding he or she lives in. The interpreter of a text may, however, attempt to make his presuppositions explicit, and attempt to become conscious of how certain formulations of a question to a text already determine which forms of answers are possible.

A seventh canon states that every interpretation involves innovation and creativity—*"Jedes Verstehen ist ein Besserverstehen"* (Every understanding is an understanding-better). The interpretation goes beyond the immediately given and enriches the understanding by bringing forth new differentiations and interrelations in the text, extending its meaning.

SOURCE: Adaptation and extension of *Contemporary Schools of Metascience* (p. 218), by G. Radnitzky, 1970, Gothenberg, Sweden: Akademiforlaget.

In Box 12.2, we have outlined some canons of interpretation developed within the hermeneutic tradition of text interpretation in the humanities that we touched upon in Chapter 3, and which we believe may serve to clarify the issues raised by multiple interpretations of interview texts. The hermeneutic principles have been sought for arriving at valid interpretations of religious, legal, and literary texts (see Palmer, 1969). It should be borne in mind that hermeneutics does not involve any step-by-step method, but is an explication of general principles found useful in a long tradition of interpreting texts, such as the canons presented in the box above. It is further an issue of debate whether hermeneutics should involve specific techniques for the interpretation of texts or whether hermeneutics is primarily a general questioning of the meaning of being. In the tradition of Gadamer (1975), it is thus explicitly rejected that hermeneutics is a method, and instead understanding is posited as the fundamental mode of being for humans. In the following sections we pursue the methodological implications of hermeneutics.

The Primacy of the Question in Interpretation

A common objection to interview analyses goes like this: "Different interpreters find different meanings in the same interview, the interview is thus not a scientific method." Dissimilar interpretations of the same interview passages do occur, though probably less often than is commonly assumed. The above objection involves a demand for objectivity in the sense that a statement has only one correct and objective meaning, and the task of an analysis is to find this one and only true meaning. Contrary to such a requirement of unequivocality, hermeneutical and postmodern thought allow for a legitimate plurality of interpretations.

There are multiple questions that can be posed to a text in an analysis, with different questions leading to different meanings. A researcher's presuppositions enter into the questions he or she poses to a text and thus codetermine the subsequent analysis. We saw above how the meaning of Susan's smile was interpreted differently with each new context, the selections of the relevant contexts rendered questions belonging to different theoretical perspectives. Some hermeneutic distinctions of types of questions to analyses of texts now follow. A first question concerns the relation of the author's and the reader's meanings. Is the purpose of a text interpretation to get at the *author's intended meaning* of the text—what Ibsen really meant to say with his play *Peer Gynt*—or does it aim to analyze the *meaning the text has for us* today? The interpretation of an interview involves a related distinction—is the purpose to analyze, for example, interviews about grades in order to arrive at the individual pupils' understanding of their grades? Or is the aim for the researcher to develop, through the pupils' descriptions, a broader interpretation of the meaning of grades in the educational system?

Another issue in interpretation concerns whether it is the *letter of the text* or *its "spirit"* that is to be interpreted in, for example, a legal text. Is what matters to get at the expressed meaning or at the intended meaning? In interview studies, this becomes a question of the level on which the interpretations should take place: Should the interviews be analyzed on a manifest level? Or is the purpose to get at latent meanings that are not explicitly conscious for the subject, as in the "depth hermeneutics" of psychoanalysis?

Interpretations of meaning are sometimes steeped in mistrust of what is said. Hamlet's interview was thus read as expressing a pervasive distrust of the words and acts of the other players, leading to a conversation of "per indirections find directions out." Within a "hermeneutics of suspicion," statements are critically interpreted as meaning something else than what is manifestly said, such as when a psychoanalytic interpreter looks for unconscious forces beneath what is said, or Marxist interpreters look for ideological class interests behind political statements.

A third issue implies the principal question of whether there exists *one correct interpretation* of a literary text or of a Bible story; or whether there is a *legitimate plurality of interpretations*. Can the gospels of the New Testament thus be said to have one correct interpretation, or are they essentially ambiguous, open to different interpretations? If the principle of a legitimate plurality of interpretations through interview analyses is accepted, it becomes meaningless to impose strict requirements of interpreter consensus. What then matters is to formulate explicitly the evidence and arguments that enter into an interpretation, in order that other readers can test the interpretation.

In current interview research, the main problem is not a lack of variety of analyses and interpretations, but rather the lack of explicit formulations of the research questions to a text. We may here distinguish between a biased and a perspectival subjectivity. A *biased subjectivity* simply means sloppy and unreliable work; researchers noticing only evidence that supports their own opinions, selectively interpreting and reporting statements justifying their own conclusions, overlooking any counterevidence. A *perspectival subjectivity* appears when researchers who adopt different perspectives and pose different questions to the same text come up with different interpretations of the meaning. When the readers' different perspectives on a text are made explicit, the different analyses should also become comprehensible. Subjectivity in this sense of multiple perspectival interpretations will then not be a weakness, but testify to the fruitfulness and the vigor of interview research.

A fourth issue involves the question of what aspects of a theme should be analyzed, and in which context. Hermeneutical text interpretations, psychoanalytical studies, and also psychological interview investigations have often involved an individualistic and idealistic focus on the intentions and experiences of individuals. There has been a neglect of the *social and material context* that the persons live in; see Sartre's critique of the "psychoanalyzing" of Robespierre's reasons for his political behavior (Box 13.2). The interview method as such does not, however, need neglect the social and economic aspects of the human situation. It is mainly the contexts in which it has been used that have given the interview research this idealistic or individualistic slant.

Analytic Questions Posed to an Interview Text

The relationship between questions to, and answers from, a text in the process of analyzing will be illustrated with an interpretation of the following interview statement about grades.

> I know that somebody will say that it is wheedling ("apple polishing") if one
> seems to be more interested in a subject matter than is usual and says: "This

is really interesting," asks a lot of questions, wanting explanations. I don't think it is . . .

In religious instruction, where we get grades (from the teacher), but do not have an examination at the end of the school year, there is plenty of time to talk about anything else. Well, people do their homework during these lessons, and then we sometimes, perhaps two or three of us, discuss something interesting with the teacher. And then, afterwards, it sometimes happens that someone remarks: "Well, well, somebody seems to be wheedling."

(*Later on in the interview, about other pupils*) Sometimes we don't know whether they do it in order to wheedle or not, but at other times it seems very opportunistic. (*In a tense voice*) It's rather unpleasant. . . . It isn't easy to figure out whether people wheedle or whether they're just interested.

This high school girl's statement is rich in information about the influence of grading on the relationships between teachers and pupils. It is, however, not quite clear what her remarks mean. One line of inquiry addresses the meaning of the text in three different interpretational contexts: self-understanding, a critical commonsense understanding, and a theoretical understanding.

Contexts of interpretation are presented in the left-hand column in Table 12.2, and in the center and right-hand columns are the corresponding communities for and forms of validation, to which we return in Chapter 15.

Self-understanding. The interpreter here formulates in a condensed form what the subjects themselves understand to be the meanings of their statements. The interpretation is more or less confined to the subjects' self-understanding in the form of a rephrased condensation of the meaning of the interviewees' statement from their own viewpoints, such as these are understood by the researcher. The meaning condensation used by Giorgi, and also the categorization of the grade interviews, took place within the context of the subject's self-understanding.

Table 12.2　　Contexts of Interpretation and Communities of Validation

Context of interpretation	Community of validation	Form of validation
Self-understanding	The interviewee	Member validation
Critical commonsense understanding	The general public	Audience validation
Theoretical understanding	The research community	Peer validation

The pupil in the statement above is interested in religion and enjoys discussing it with the teacher, but she has the impression that other pupils may regard this as wheedling. In other situations, she has difficulties determining whether the other pupils wheedle or whether they are actually interested in the subject matter. She experiences this ambiguity as rather unpleasant.

Critical Commonsense Understanding. The interpretation here goes beyond reformulating the subjects' self-understanding—what they themselves experience and mean about a topic—while remaining within the context of a commonsense understanding. The analysis may thus include a wider frame of understanding than that of the subjects themselves, may be critical of what is said, and may focus on either the content of the statement or on the person making it. The interpretation of the statement with the denials of competition mentioned earlier (Chapter 10) thus went beyond the pupil's self-understanding to include a critical commonsense reading of the many denials as possibly indicating a confirmation.

By including general knowledge about the *content* of the statement it is possible to amplify and enrich the interpretation of a statement. For the question "What does the statement express about the phenomenon of wheedling?" the girl's statement may be interpreted as a manifestation of a basic ambiguity in the teacher-pupil relationship created by grading. Within a grade-dominant perspective, the subject matter and the human relationships in school are "instrumentalized": They become mere means toward the goal of the highest possible grade point average. In the classroom it may appear ambiguous whether a pupil's expressed interest in a topic is genuine, or whether it is just a means to "twist" the teacher in the interest of improving grades.

The questions put to the text may also center on the *person,* asking what a statement expresses about the interviewed subject. In the statement above, the question "What does it express about the pupil's own relation to wheedling?" may lead to an interpretation that this girl employs double standards: The same activity of talking interestedly with the teacher is evaluated more positively when conducted by the girl herself than when carried out by others. The topic involves a conflict for her; her voice is tense, and a speculative interpretation might be that she belongs to that group of pupils whom the others accuse of wheedling.

Theoretical Understanding. In a third context, a theoretical frame for interpreting the meaning of a statement is applied. The interpretations are then likely to go beyond the subject's self-understanding and also to exceed a commonsense understanding, such as when incorporating a psychoanalytic theory of the individual or a Marxist theory of society.

In a somewhat speculative interpretation, the psychoanalytical concept of "projection" may be used: At an unconscious level the pupil projects her own nonacceptable wheedling behavior onto other pupils, while denying it in herself. In a Marxist theory about the school system as socializing to wage labor, with grades as the currency of the school system (Bowles & Gintis, 1976; Kvale, 1972), the statement about apple polishing may be interpreted as an expression of learning at school having a "commodity character." The pupils learn—through the grading of their learning—how to distinguish between the use value and the exchange value of their work. Their questions to the teacher may be led by a utility interest in obtaining a better understanding of the knowledge presented. The questions may also be part of an instrumental exchange relation; the knowledge about which they ask interested questions has no intrinsic use value for the pupils, and the questions only serve the purpose of making a positive impression on the teacher—an impression that can be exchanged for a higher grade. At school the pupils thus learn to subordinate the use value of their work to its exchange value.

Interrelatedness of Interpretational Contexts. The three interpretational contexts for analysis derive from different explications of the researcher's perspective and lead to different forms of analysis. The contexts may be further differentiated, and they may also merge into each other. The instrumental attitude toward learning—knowledge as a mere means to high grades—which was discussed above in a commonsense context, also follows from sociological and Marxist theories about education. For some of the Danish pupils in the study, such an instrumental means-ends thinking was an open part of their self-understanding: "My interests have taken me very far from that which takes place at high school. I go here with the explicit purpose of getting as good an examination as possible, with the least possible effort." The three contexts of interpretation suggested above serve to make explicit the analytic questions posed to a statement. One pupil's description of wheedling has given rise to a number of interpretations in our analysis. The various resulting interpretations are, according to the present perspective, not haphazard or subjective, but follow as answers to different questions to the text.

Implications for Interviewing and Transcription: For extensive interpretations of meaning, rich and nuanced descriptions in the interviews are advantageous, as are critical interpretative questions during the interview. For some types of interpretation, detailed verbatim descriptions may be necessary, such as when critically reading a pupil's many denials of competition in Chapter 10.

THE QUEST FOR "THE REAL MEANING"

When analyzing the meanings of an interview, a common question asked of interview researchers goes something like "How do you know you get to know what the person really means?" A tempting reply: "What do you really mean by 'really means'?" will probably not lead anywhere.

Guessing at the meaning of "the 'real meaning' question" suggests a belief in the existence of some basic meaning nuggets stored somewhere, to be discovered and uncovered, uncontaminated, by the objective techniques of an interviewer understood as a miner digging up precious buried metals. The "real meaning" question is a leading question, in this case leading to endless pursuits of an undefined and fictitious entity. The quest for real, true meanings came to an end in philosophy some years ago. Interview researchers might still go on wild goose chases, hunting the real, authentic, meanings of their subjects' experiences. Psychotherapists might still be digging for real meanings in the deep interior of their patients' unconscious psyches. Both therapist and patient then conceive of truth as something to be found or "mined," and not as something socially constructed with their subjects.

The question of real meanings raises the issue of who owns the meanings of a statement. A constructed dialogue can illustrate the issue of ownership of meanings:

A: Did you really mean that?

B: No, that is not what I said.

A: Oh yes, you said it and you did mean it!

B: I know what I wanted to say, and I know that I did not mean what you say I meant!

A: I know you, and I know what you really mean!

In this interchange, two themes are disputed: the true meaning of a statement is explicitly disputed, and, somewhat more implicitly, who has the right and the power to determine the real meaning of the statement—the speaker of the original statement or the interpreting partner. An interrelational interpretation would regard the meanings of the conversation as belonging to neither, but existing between the subjects, in their inter-action. The conversation is then seen as interchanges of a power game, a contest for who in the relationship possesses the right to attribute the definite meaning to a statement. The search for real-meaning nuggets may lead to reification of the subjective rather than to an unfolding and an enrichment of the subjective, which follows from an interrelational conception in which meanings are constructed and reconstructed through conversational interactions. A postmodern approach laid out in Chapter 3 forgoes the search for true fixed meanings and emphasizes descriptive nuances, differences, and paradoxes. There is a change from a substantial to a relational concept of meaning, with a move from a modernist search for the one true and real meaning to a relational unfolding of meanings. That different interpreters construct different meanings of an interview story is then not a problem, but a fruitfulness and virtue of interview research.

With a transition from an individual storage conception of meaning to an interrelational constitution of meaning in the original interview conversation—and in the readers' conversations with the interview text—the power asymmetry of the interview researcher and the subject become more obvious. Does the interviewer own the meanings constructed in and on an interview, interpreting it within his or her selected contexts? Or should the original "authors" of the interview statements have their say in the interpretation and communication of their stories? This is an issue not only of validity of interpretation, but of ethics and power, of the right and the power to attribute meaning to the statements of others.

In the meaning dialogue above, the partners appeared to be on an equal social level, while contesting who was in power. If the interpreter had the status of a professional expert, the original speaker might more humbly have accepted the "real" meanings attributed to him or her. This can be meaningful and legitimate as new stories told by the interpreter, but if reified as the real meaning of the interview subject, or as the real unconscious meaning of the patient, they become more problematic ethically and with respect to validity. It should not be overlooked that the implicit, or unconscious, meanings attributed to interviewees may often simply be the explicit and conscious theories of the expert interpreter. Interview research involves the danger of an "expertification" of meanings where the interviewer as "the great interpreter" expropriates the meanings from the subjects' lived world and reifies them into his or her theoretical schemes as expressions of some more basic reality.

THIRTEEN

INTERVIEW ANALYSES FOCUSING ON LANGUAGE

————⊷•◆•⊶————

The interview process occurs through speech, and the interview products are presented in words. The medium, or the material, with which interviewers work is language. During the last few decades, qualitative social science researchers have been influenced by the linguistic turn in philosophy, and they have started to apply linguistic tools developed in the humanities to analyze their linguistic material. These include linguistic analysis, conversation analysis, narrative analysis, discourse analysis, and deconstruction, which we will now exemplify and discuss in relation to analyses of interviews. While linguistic and conversation analysis mainly focuses on the linguistic structures, the latter forms of analysis address both linguistic structures and meaning. Here we will explore how the different modes of analysis entail different conceptions of the meanings and language of interview texts and how they lead to different questions to the analyzed material.

LINGUISTIC ANALYSIS

Interviewing is linguistic interaction, and the product of the interview is a language text. A linguistic analysis addresses the characteristic uses of language in an interview, the use of grammar and linguistic forms. A linguistic analysis may thus study an interviewee's use of active and passive voice and personal and impersonal pronouns, temporal and spatial references, the implied speaker and listener positions, and the use of metaphors.

An example from the grading study may indicate the importance of linguistic form. The analysis did not follow from any linguistic competence of the researcher, but arose as a practical problem of categorizing the pupils' statements. While most grading experiences and behaviors were commonly described in a first-person form, such as "I find the grades unfair" and "I bluffed the teacher," a few activities, such as wheedling, were always described in a third-person form, such as "They wheedled" or "One wheedled." If the researchers had then been more sensitive to the differential use of personal pronouns, we might, in the interviews, have probed more into such a vague expression as "one wheedles" and clarified whether it referred to the speaker or to other pupils. While the ambiguous use of personal pronouns was a method problem when categorizing the statements as referring to the interviewee or to the other pupils, it was of importance to the research topic as one of the many indications of the contrasting social acceptance of bluffing versus wheedling among the Danish high school pupils.

Attention to the linguistic features of an interview may contribute to both generating and verifying the meaning of statements. While understanding the significance of the different uses of grammatical forms such as the above example of personal pronouns may follow from common sense, a linguistically trained reader would immediately look for the linguistic expressions and be able to bring out nuances, which may be important for interpreting the meaning of a statement. Arguments in favor of applying the techniques of linguistics as a "statistics" of qualitative research have even been put forward (Jensen, 1989). With more attention to the linguistic medium of interview research, we may perhaps see social researchers use linguists as consultants when faced with interview texts, corresponding to the commonplace use of statistical consultants when analyzing quantitative data.

Implications for Interviewing and Transcription: Attention to linguistic form may improve the preciseness of interview questions, and further sensitivity in listening to the subjects' use of language. To carry out systematic linguistic analyses of the interview interaction, detailed verbatim transcription and also linguistic training is necessary.

CONVERSATION ANALYSIS

Conversation analysis is a method for studying talk in interaction. It investigates the structure and the process of linguistic interaction whereby intersubjective understanding is created and maintained. Inspired by ethnomethodology (Garfinkel, 1967), conversation analysis implies a pragmatic theory of language, it is about what words and sentences do, and the meaning of a statement is the role it plays in a specific social practice. Conversation analysis started with studies of telephone conversations by Sacks and his colleagues in the 1960s, and has since been used for a wide variety of talk in action, such as doctor-patient interactions, therapy sessions, and news interviews. This methodical "conversation analysis" differs from the use of "conversational analysis" in a more general sense for reading a text by Bourdieu (see Box 6.3) and others.

Conversation analysis examines the minute details of talk-in-interaction, which became generally accessible with the advent of tape recorders. It focuses on the sequencing of talk, in particular upon turn-taking sequences and repair of turn-taking errors. The center of attention is not the speakers' intentions in a statement, but what a specific speech segment accomplishes. Consequently, the outcome of the conversation analysis of the telephone conversation transcribed in Box 10.1 was:

> E apparently has called M after having visited her. She provides a series of "assessments" of the occasion, and M's friends who were present. E's assessments are relatively intense and produced in a sort of staccato manner. The first two, on the occasion and the friends in general are accepted with Oh-prefaced short utterances, cut-off when E continues . . . The assessments of Pat are endorsed by M with "yeh," followed by a somewhat lower level assessment. "a do:11?" with "Yeh isn't she pretty" and "Oh: she's a beautiful girl.," with "Yeh I think she's a pretty girl." . . . The "work" that is done with these assessments and receipts can be glossed as "showing and receiving gratitude and appreciation, gracefully." (ten Have, 1999, p. 4)

Conversation analysis thus sticks rather close to the verbal interaction of the speakers, forgoing interpretations in depth. The labor-intensive transcription and the minute analyses of the speech sequences rule out conversation analyses as a general method for analysis of large amounts of interview material. Conversation analysis may, however, be relevant for selected significant parts of an interview, and it may also be useful in the training of interviewers, to make them aware of the subtleties of the interaction in the interviews. Some researchers within the field of discourse analysis are skeptical of the merits of conversation analysis. Parker (2005, p. 91) thus criticizes its "textual empiricism"— that you can only address what you can see directly in the transcript, thus excluding power and ideology, for example, if they are not explicitly mentioned— and also what he sees as "pointless redescription," where researchers simply repeat what is said in the transcript in a more detailed way, clouded in seemingly rigorous conversation analysis jargon.

Implications for Transcription: Here there are no specific requirements for interviewers since any verbal exchange can be made the subject of conversation analysis. As was seen in the transcription in Box 10.1, there are, nevertheless, very specific and elaborate requirements for how interviews are to be transcribed in order to be amenable to conversation analysis.

NARRATIVE ANALYSIS

A narrative is a story. Narrative analyses focus on the meaning and the linguistic form of texts; they address the temporal and social structures and the plots of interview stories. The narrative structures of stories people tell have been worked out in the humanities, starting with Propp's analysis of the structures of Russian fairy tales in the 1920s, and followed up decades later by Greimas and Labov. In the structure of a fairy tale, the main subject position may be taken by the prince as the protagonist, who seeks the object in the form of the princess. On his way, the prince encounters opponents as well as helpers, and after overcoming the many obstacles, the prince receives from the king the princess and half his kingdom. Greimas used this structure to work out an actant model pertaining to narrative structures in a variety of genres.

Narrative analysis focuses on the stories told during an interview and works out their structures and their plots. If no stories are told spontaneously, a coherent narrative may be constructed from the many episodes spread throughout an interview. The analysis may also be a reconstruction of the many tales told by the different subjects into a "typical" narrative as a richer, more condensed and coherent story than the scattered stories of single interviews. As with meaning condensation, narrative analysis will tend to stay within the vernacular.

A narrative sequence from Mishler's interview with a furniture craftsman-artist dropping out of the woodworking program at school was presented in Box 8.3. We may also note that the interview on learning interior decorating from the previous chapter had the spontaneous form of a narrative, which Giorgi did not address in his meaning condensation of the interview. Chapter 6 on designing an interview study was introduced by a constructed narrative of successive emotional deterioration, followed by an idealized story of a linear progression through seven stages of an interview inquiry.

A narrative can be analyzed in many different ways, and here we will mention analyses of *structure, plot,* and *genre.* In a linguistic sense, a narrative analysis concerns a chronologically told story, with a focus on how its elements are sequenced. Labov has put forth a standard linguistic framework for narrative analyses (Labov, 1972; Labov & Waletzky, 1967). Using this framework in an analysis can help highlight the structure of a narrative, by breaking the narrative down into specific interconnected components. These are (1) the *abstract* that provides a summary of the narrative, inserting it into the framework of conversational turn-taking; (2) the *orientation* of the listener to the time, place, actors, and activity of the narrative; (3) the *complicating action,* containing the central details of the narrative; (4) the *evaluation* of the central details; and (5) the *coda,* which summarizes and returns the narrative time frame to the present (Murray, 2003).

The first four of these components are illustrated in Box 13.1, which contains a narrative Labov calls "The first man killed by a car in this town," and is based on an interview Labov conducted in 1973 with a 73-year-old man from South Lyons, Michigan.

BOX 13.1   The First Man Killed by a Car in This Town: A Narrative

Abstract

Shall I tell you about the first man got kilt–killed by a car here . . .

Well, I can tell you that.

Orientation

He–eh–'fore-'fore they really had cars in town

I think it was a judge–Sawyer–it was a judge in–uhc

I understand he was a judge in Ann Arbor

and he had a son that was a lawyer.

(Continued)

(Continued)

Complicating action

And this son–I guess he must've got drunk

    because he drove through town with a chauffeur

    with one of those old touring cars without, you know–

    open tops and everything, *big* cars, first ones–

and they–they come thr-through town in a–late in the night.

And they went pretty fast, I guess,

and they come out here to the end of a–

    where–uh–Pontiac Trail turns right or left in the road

and they couldn't make the turn

and they turned left

and they tipped over in the ditch,

steerin' wheel hit this fellow in the heart, this chauffeur,

killed him.

Evaluation

And–uh–the other fellow just broke his thumb–

    the lawyer who [hh] was drunk.

They–they say a drunk man [laughs] never gets it [laugh].

Maybe I shouldn't say that,

I might get in trouble.

SOURCE: Adapted from *Uncovering the Event Structure of a Narrative* (pp. 3–4), by W. Labov, March 2001, a paper given at the Georgetown Round Table. Retrieved May 24, 2007, from http://www.ling.upenn.edu/~wlabov/uesn.pdf

The sequencing in Box 13.1 represents a structural analysis, but Labov also supplies us with an analysis of the meaning or the *plot* of the narrative:

> We have no difficulty in understanding this narrative in terms of its main point, established in the evaluation section [. . .]. Big city lawyers are the problem, and the blame is clearly assigned to the drunken lawyer, who

escaped with minor injuries. The narrator makes a little joke about the fact that he, a small town person, might get into trouble by criticizing these city folks with their chauffeurs and big, modern, fancy touring cars. It is the most familiar theme of American culture: the simplicity, honesty and competence of small town people against the sophistication, corruption and incompetence of big city people. (Labov, 2001, p. 4)

Narrative researchers have tried to identify the plots that typically appear in stories. Plots are what order narratives, and the basic plots of modernist stories have been identified as taking a journey, engaging in a contest, enduring suffering, pursuing consummation, and establishing a home. In addition to analyses of structure and plot, narrative analyses can also aim at ascertaining the *genre* of a story told. The classical literary model thus distinguishes the four genres of comedy, romance, tragedy, and satire (see Murray, 2003, for an overview).

As with any form of qualitative analysis, the narrative approach has its limitations. Recently, Frosh (2007) criticized "narrativism" from a psychoanalytic perspective, arguing that not all aspects of human experience fit neatly into more or less coherent narratives, for human subjects are not just integrated through narratives, but also fragmented. He defends the idea that the human subject is *"never* a whole, is always riven with partial drives, social discourses that frame available modes of experience, ways of being that are contradictory and reflect the shifting allegiances of power as they play across the body and the mind" (p. 638). Frosh does not want us to abandon narrative analysis, but rather wants qualitative researchers to hold on to the dialectic of deconstructivist fragmentation and narrative integration when describing human experience. Qualitative research exists, as he says, in a tension between "on the one hand, a deconstructionist framework in which the human subject is understood as positioned in and through competing discourses and, on the other, a humanistic framework in which the integrity of the subject is taken to be both a starting- and end-point of analysis." (p. 639).

Implications for Interviewing and Transcription: Interviewing for narratives was described in Chapter 8, where questioning for concrete episodes and the following up of the subjects' spontaneous stories, elaborating their temporal and social structures and plots, were emphasized. When transcribing, one may experiment with the textual layout in ways to make the narrative form accessible, such as with the stanzas in Mishler's craftsman story and the subheadings in Labov's car story.

DISCOURSE ANALYSIS

Discourse analysis focuses on how truth effects are created within discourses that are neither true nor false. Foucault's (e.g., 1972) analyses of the power relations of discourses have inspired later forms of discourse analysis. Parker defines discourse as *"the organization of language into certain kinds of social bonds"* (2005, p. 88). In Parker's version of discourse theory, the approach becomes closely aligned with the study of ideology. An ideology is an organized collection of ideas, and discourse analysis attempts to unravel how such collections of ideas are produced and work in practice. Discourse analysis studies how language is used to create, maintain, and destroy different social bonds, and is in line with the postmodern perspective on the human world as socially and linguistically constructed, which we outlined in Chapter 3. It shares with pragmatism an emphasis on the primacy of doing, of practice, of actions performed in the here and now. Discourses are discontinuous practices, which cross each other and sometimes touch, while just as often ignoring or excluding each other. In Chapter 8, an interview sequence was presented, which, inspired by discourse analysis, was analyzed as a crossing of swords of the diverging discourses of learning of the interviewer and the electronics pupil.

In some respects, the currently popular discursive perspective comes close to a dialectical emphasis on contradictions such as that found in Marxist philosophy (Parker, 2005). Here we will outline some aspects of the multifaceted line of dialectical thought.

BOX 13.2   Discourse and Dialectics

Dialectics is the study of internal contradictions—the contradiction between the general and the specific, between appearance and essence, between the quantitative and the qualitative, for example. The development of contradictions is the driving force of change, according to dialectical philosophy.

Dialectical materialism involves the fundamental assumption that the contradictions of material and economic life are the basis of social relations and of consciousness. Human consciousness and behavior are studied within the concrete sociohistorical situation of a class society and its forces and relations of production. The objects of the human sciences are seen as multifaceted and contradictory, consisting of internally related opposites in continual change and development. Human beings act upon the world, change it, and are again changed by the consequences of their actions.

When doing interviews from a discursive or dialectical viewpoint, one is interested in the contradictions that individuals articulate not as aspects of concrete individuals per se, but rather as aspects of historical discursive practices. Sartre outlined this in his attempt to mediate between Marxism, existentialism, and psychoanalysis in *The Problem of Method* (1963). His critique of the individualizing approach of psychoanalysis also pertains to much current individualist interview research: "How many times has someone attempted the feat of psychoanalyzing Robespierre for us without even understanding that the contradictions in his behavior were conditioned by the objective contradictions of the situation" (p. 60). Sartre's parallel critique of an objectifying Marxist reductionism might also be mentioned: "Valéry is a petit bourgeois intellectual, no doubt about it. But not every petit bourgeois intellectual is Valéry" (p. 56).

Discursive and dialectical approaches are not identical, for the former typically involves a constructionist perspective, whereas the latter approach is more objectivist and realist. In both the discursive and the dialectical approaches, however, there is an emphasis upon contradictions and the new, upon what is becoming and under development. With a conception of the social world as being developed through contradictions, discursively and materially, it is important to uncover the new developmental tendencies in order to obtain true knowledge of the social world. The statistical average, or the representative case of the status quo, are less important than the new tendencies developing as the *status nascendi*. If social reality is in itself contradictory, the task of social science is to investigate the real contradictions of the social situation and posit them against each other. In other words, if social processes are essentially contradictory, then empirical methods based on an exclusion of contradictions will be invalid for uncovering a contradictory social reality.

In Box 13.2, we have attempted to demonstrate the importance of contradictions in studies of discourse—something they share with a dialectical approach. In discourse analysis, the talk itself has primacy:

Participant's discourse or social texts are approached in *their own right* and not as a secondary route "beyond" the text like attitudes, events or cognitive processes. Discourse is treated as a potent, action-oriented medium, not a transparent information channel.

Crucial questions for traditional social psychological research thus cease to be relevant. For example, we are not asking whether a sample of people are revealing their "genuine" attitudes to ethnic minorities, or whether fans' descriptions of what happens on the soccer terraces are "accurate." (Potter & Wetherell, 1987, p. 160)

From a discourse-analytic perspective, some common objections to the validity of research interviewing thus dissolve. This concerns the question of authentic personal meanings—"How do you know you get to know what the interviewees really mean?"—as well as the objective reality question—"How do you know that your interviewee gives a true description of the objective situation?" A persistent objection concerning the reliability of interviewing has been that different interviewers get different results. If subjects present themselves differently to different interviewers, and also change their opinions during the interchange, then interviews do not produce reliable, objective knowledge.

These objections may be based upon conceptions of the research topic, such as attitudes or presentation of the self, as expressions of an essential stable core person. In contrast, a discursive understanding treats attitudes and the self as interrelationally constituted, as emerging out of discursive acts and performances in social interaction. These phenomena may vary in different situations with different interviewers, and interviewing is a sensitive method to investigate the varying social presentations of the self. Thus, according to the differing epistemological conceptions of attitudes and the self—as stable authentic essences, or as socially constituted and more or less fluid ones—the interview appears either as a highly unreliable or as a finely tuned and valid method.

A clear example of discourse analysis is found in Wetherell and Potter's (1992) research on white New Zealanders' (Pakeha) construction of "culture," "race," and "nation" through their use of discourses or what the authors call interpretative repertoires. The authors conclude that even seemingly egalitarian discourses can contribute to increase racism and discrimination. Wetherell and Potter discovered a general discursive change from the 70s race discourse to contemporary society's cultural discourse. They identified different cultural discourses, a significant one being culture-as-inheritance, which is seen in the following example, where culture is constructed as unchanging and traditional:

Williamson: I think it's important they hang on to their culture (yeah) because if I try to think about it, the Pakeha New Zealander hasn't got a culture (yeah).

I, as far as I know, he hasn't got one (yeah) unless it's rugby, racing and beer, that would be his lot (yes) But the Maoris have definily got something, you know, some definite things that they do and (yeah). No, I say hang onto their culture. (Wetherell & Potter, 1992, p. 129)

Potter and Wetherell interpret this and other interview passages to mean that the Maoris are constructed as "museum attendants," obliged to maintain their culture for their own sake, which is a construction that has ideological effects by separating culture and politics.

There are many different approaches to discourse analysis, some of which are formal and technical, and others less prescribed. Parker (2005, p. 92) advocates a nonformal approach based on four questions that the researcher can pose to a text when engaging in discourse analytical work: Why is the text interesting? What do we know of the material out of which it was constructed? What are the effects of different readings of the text? How does the text relate to patterns of power? We can illustrate the significance of the questions by looking at a short example, in this case a joke:

Mrs Thatcher goes to a restaurant with her cabinet. She tells the waiter that she will have chicken. The waiter says, "What about the vegetables?" and she replies, "They'll have chicken too." (Parker, 2005, p. 93)

Summarizing Parker's analysis, we may ask first, why is the text interesting? In general, it is not coincidental which parts of a text one stumbles upon as calling for analysis. Qualitative researchers normally choose, and often with good reasons, to analyze those aspects of their material that challenge their preconceptions, seem puzzling in relation to the research question, or simply stand out one way or the other. The Thatcher joke above contains a puzzle that deserves further analysis concerning why it is funny. Second, the researcher should find out what he or she knows of the material out of which the text was constructed. In the case of the joke, it is relevant to consider which political images and ideas lurk in the background and give the joke its sense and point. In the joke, a number of discourses can be identified, such as discourses of masculinity (the members of the cabinet are "less than men") and of women's supremacy (perhaps the joke is funny because it reverses the traditional discursive order of male dominance).

Third, the researcher can ask what the effects might be of different discursive readings of a text. The joke can be read from a feminist standpoint, stressing the strength of Thatcher as an assertive political figure who jokes about the male members of her cabinet, or it can be read as a way of disclosing Thatcher as a caricature, a figure of fun, "handbagging" her opponents. Different readings are possible, and these are themselves forms of social action that will have different discursive effects in practice. Finally, a discourse analyst can ask how the text conforms to or challenges patterns of power. If speech and action are discursive performances that either work for or against power, the question of power is always relevant from a discourse theoretical point of view. How are speakers positioned in social networks of power that grant different rights and duties to speak and act in certain ways? Positioning theory is one recent variety of discourse analysis that highlights the issue of social positioning (see Harré & Moghaddam, 2003), and in the joke example we clearly see positioning in play in the unequal power relationship between Thatcher and her cabinet.

Implications for Interviewing: While discourse analysis may be applied to common interviews, such as the one described as "discourses crossing swords" (see Box 8.4), a specific discursive interviewing will focus on variation and diversity, and on the active participation of the interviewer in the discourse, as shown in Chapter 8. The search for real inner meanings and objective presentations of external reality dissolve, and instead there is a focus on the discursive production of a social episode.

DECONSTRUCTION

Derrida introduced the concept of "deconstruction" as a combination of "destruction" and "construction." Deconstruction involves destructing one understanding of a text and opening for construction of other understandings (see Norris, 1987). The focus is not on what the person who uses a concept means, but on what the concept says and does not say. It is affiliated with a critical "hermeneutics of suspicion," but in line with conversation and discursive analyses, it does not search for any underlying genuine or stable meaning hidden beneath a text. Meaning is understood in relation to an infinite network of other words in a language.

A deconstructive reading tears a text apart, unsettles the concepts it takes for granted, concentrates on the tensions and breaks in a text, on what a text purports to say and what it comes to say, as well as what is not said in the text, on what is excluded by the use of the text's concepts. A deconstructive reading reveals the presuppositions and internal hierarchies of a text and lays open the binary oppositions built into modern thought and language, such as true/false, real/unreal, subjective/objective. Deconstruction does not only decompose a text, but also leads to a redescription of the text.

A deconstructive reading could, for example, focus on selected interview passages and phrases and work out the meanings expressed, as well as meanings concealed and excluded by the terms chosen. Rather than deconstructing an interview text, we shall here attempt to deconstruct a phrase recurrently evoked in interview literature, and in the first edition of the present book—"the interview dialogue."

We may start by wondering why the two similar terms *interview* and *dialogue* are often added together, and not uncommonly bolstered with embellishing words, such as *authentic, real, genuine, egalitarian, and trusting. Dialogue* exists in a binary opposition to *monologue,* which today may connote an old-fashioned, authoritarian form of communication. When dialogue is used in current interview research it is seldom in the rigorous conceptual Socratic form, but more commonly as warm and caring dialogue. "Dialogical interviewing" can suggest a warm empathetic caring, in contrast to alienated and objectifying forms of social research, such as those found in experiments and questionnaires. When the interview is conceived as a dialogue, the implication is that the researcher and subject are egalitarian partners in a close, mutually beneficial personal relationship. The expression "interview dialogue" here glosses over the asymmetrical power relationships of the interview interaction, where the interviewer initiates and terminates the interview, poses the questions, and usually retains a monopoly of interpreting the meaning of what the interviewer says (see Box 2.3).

We further note that the term *dialogue* is used today in texts from a variety of fields, such as management and education, which advocate "dialogue between managers and workers" and "dialogical education" (Kvale, 2006). In these contexts, with obvious power differences and often conflicts, the term *dialogue* may provide an impression of equality and harmonious consensus. We will conclude this brief deconstruction of the phrase "interview dialogue" by asking whether the term *dialogical interviews* used about research interviews, may, corresponding to dialogical management and dialogical education, serve to embellish the power asymmetry and cover up potential conflicts of research interviewers and their interviewees.

Implications for Interviewing: As any kind of text may be made a subject of deconstruction, there are no specific requirements for interviewing. If a deconstruction of interview texts is considered, the interviewer may, however, address the use of key terms in a variety of contexts, from a multiplicity of perspectives, thereby providing multifaceted material for deconstruction. Although there is, to our knowledge, no worked out practice of deconstructivist interviewing, some of Garfinkel's (1967) ethnomethodological breaching experiments approach a deconstructive interview. In some experiments, Garfinkel asked his students to engage in ordinary conversations and insisted that the person clarify the sense of commonplace remarks:

(S)   Hi, Ray. How is your girl friend feeling?

(E)   What do you mean, "How is she feeling?" Do you mean physical or mental?

(S)   I mean how is she feeling? What's the matter with you? (He looked peeved.)

(E)   Nothing. Just explain a little clearer what do you mean?

(S)   Skip it. How are your Med School applications coming?

(E)   What do you mean, "How are they?"

(S)   You know what I mean.

(E)   I really don't.

(S)   What's the matter with you? Are you sick? (Garfinkel, 1967, pp. 42–43).

"S" is the subject and "E" is the "experimenter," deliberately breaching the ordinary production of social order. Like deconstructive analyses that unsettle the taken for granted and reveal textual presuppositions, such conversational experiments highlight the background expectancies that are rarely thematized in everyday life, and the typical result of Garfinkel's breaches was bewilderment, unease, anxiety, and even anger on behalf of the conversationalists. We do not wish to recommend "deconstructivist interviewing" as a new method, but a deconstructive stance toward interviews can help researchers become aware of the inbuilt and largely implicit premises in the social practice of interviewing and how these influence the knowledge that is produced.

⊰ FOURTEEN ⊱

ECLECTIC AND THEORETICAL ANALYSES OF INTERVIEWS

————————

We end the presentation of interview analyses by turning from specific analytical tools to more general approaches to interview analysis. Many analyses of interviews are conducted without following any specific analytic technique. Some go beyond reliance on a single mode of analysis to include a free mixture of methods and techniques. Other interview analyses do not apply specific analytic procedures, but rest on a general reading of the interview texts with theoretically informed interpretations. Knowledge of the subject matter of analysis here carries more weight than the application of specific analytical techniques.

INTERVIEW ANALYSIS AS BRICOLAGE

Bricolage is something put together using whatever tools happen to be available, even if the tools were not designed for the task at hand. The bricolage interpreter adapts mixed technical discourses, moving freely between different analytic techniques and concepts. This eclectic form of generating meaning—through a multiplicity of ad hoc methods and conceptual approaches—is a common mode of interview analysis, contrasting with more systematic analytic modes and techniques such as categorization and conversation analysis.

The interviewer craftsman may read through the interviews and get an overall impression, then go back to specific interesting passages, perhaps count statements indicating different attitudes to a phenomenon, cast parts of the interview into a narrative, work out metaphors to capture key understandings, attempt to visualize findings in flow diagrams, and so on. Such tactics of meaning generation may, for interviews lacking an overall sense at the first reading, bring out connections and structures significant to a research project. The outcome of this form of meaning generation can be in words, in numbers, in figures and flow charts, or in a combination of these.

BOX 14.1   Ad Hoc Techniques of Interview Analysis

Noting patterns, themes (1), *seeing plausibility* (2), and *clustering* (3) help the analyst see "what goes with what." *Making metaphors* (4), like the preceding three tactics, is a way to achieve more integration among diverse pieces of data. *Counting* (5) is also a familiar way to see "what's there."

Making contrasts/comparisons (6) is a pervasive tactic that sharpens understanding. Differentiation sometimes is needed, too, as in *partitioning variables* (7).

We also need tactics for seeing things and their relationships more abstractly. These include *subsuming particulars under the general* (8); *factoring* (9), an analogue to a familiar quantitative technique; *noting relations between variables* (10); and *finding intervening variables* (11).

Finally, how can we systematically assemble a coherent understandable of data? The tactics discussed are *building a logical chain of evidence* (12) and *making conceptual/theoretical coherence* (13).

SOURCE: *Qualitative Data Analysis: An Expanded Sourcebook* (pp. 245–246), by M. B. Miles & A. M. Huberman, 1994, Thousand Oaks, CA: Sage.

Box 14.1 contains a summary of ad hoc methods from the book *Qualitative Data Analysis* by Miles and Huberman (1994), who also outline a variety of more systematic analytic techniques. In line with a bricolage approach, they present some useful ad hoc tactics for generating meaning in qualitative texts, arranged roughly from the descriptive to the explanatory, and from the concrete to the more conceptual and abstract.

In the grade study, a bricolage of mixed methods was applied to pursue a connection between talkativity and grades, postulated by a pupil (Chapter 1).

According to a questionnaire follow up (Table 6.1), 82% of the pupils believed that high grades were often a question of how much one talks in class. When reading through the 30 interviews with pupils, it was striking how significantly they varied in length, even though one school hour had been set aside for each interview. Following a hunch, the interviews were ranked according to number of pages, and then the page numbers were correlated with the pupils' grade point averages. The resulting correlation was 0.65, with a chance probability of $p < .001$. There is thus a statistically significant connection between how much the pupils talked during the interviews and their grade point averages. The correlation is open to several interpretations: Do the pupils get high grades because they generally talk a great deal? Or are pupils who get high grades more reflective about grading, and more at ease with talking at length with an interviewer about grades?

The example demonstrates how it is possible to use a variety of techniques to investigate a hypothesis of a connection between grades and talkativity: by discussing the truth value of a pupil's interview statement in the interview itself, testing the generality of the belief in a questionnaire, and finding potential indirect statistical evidence in the length of interviews about grades. In this bricolage of mixed methods there is no epistemological primacy accorded to any of the methods and techniques, they are different means of investigating a provoking statement about grading.

Implications for Interviewing: With no specific mode of analysis planned, there are no particular requirements for interviewing and transcription. The general requirement of providing rich descriptions and well-controlled information still pertains.

INTERVIEW ANALYSIS AS THEORETICAL READING

By *theoretical reading* we refer to a theoretically informed reading of interviews. In the discussion of the class interview about grading in Chapter 7, selected statements were interpreted from different theoretical positions such as Freudian, Rogerian, and Skinnerian approaches, with each approach highlighting different aspects of the student's relation to grades. In the case of Susan's smile (Box 12.1), therapists with different theoretical perspectives chose different contexts for their differing interpretations of the family interaction. And through the interpretation of the meaning of a pupil's statement about wheedling in Chapter 12, some rather speculative interpretations from psychoanalytic and Marxist perspectives were put forth. No specific methods or techniques were applied here.

A researcher may read through his or her interviews again and again, reflect theoretically on specific themes of interest, write out interpretations, and not follow any systematic method or combination of techniques. In several influential interview studies of the last decades, no systematic analytic techniques seem to have been used to analyze the interviews. This includes the work of Bellah and colleagues, Bourdieu and colleagues, and Sennett, which we mentioned earlier. These investigations were based on an extensive and theoretical knowledge of the subject matter, and in the studies by Bourdieu and Bellah, also on a confronting, Socratic interview form. No elaborate analytic techniques were applied during the theoretical reading of the interviews to develop their rich meanings. This may perhaps suggest that recourse to specific analytic tools becomes less important with a theoretical knowledge of the subject matter of an investigation, and with a theoretically informed interview questioning.

From a postmodern perspective, Lather (1995) has discussed the interrelational construction of meaning during the reading of texts. We read within a range of conventions, and Lather addresses the question of how we can learn to read our own ways of reading. Rejecting any simple analytical frame, her goal is to proliferate, juxtapose, and create disjunctions among different ways of reading, working toward a multilayered data analysis. Inspired by van Maanen's (1988) accounts of different ethnographic genres in *Tales of the Field* (see Chapter 16), Lather outlines different readings of the same text. Although her portrayal of reading styles pertains to a textbook, the styles may be transposed to the reading of interview texts. In a *realist* reading there is a search for the "native's" point of view and for finding the text's essence and truth. The reader assumes an observational and descriptive role, adopting a "god's eye point of view." A *critical* reading demystifies via a hermeneutics of suspicion; it seeks deeper truth underlying the hegemonic discourse of the texts. The reader assumes the role of the emancipator of self and/or other, seeking a truth beyond ideologies and false consciousness. The reader calls attention to larger social, political, and economical issues, assuming an advocatory role, with the danger of attempting to speak for others, of saying what they want and need. A *deconstructive* reading proliferates, destabilizes, and denaturalizes. The text is read as documentation for its unconscious silences and unspoken assumptions. A deconstructive reading makes use of drawing, artistry, literary practices, and blurs the fact/fiction distinction. These different readings suggested by Lather (1995) involve different questions posed to the text and lead to different answers about the meaning of the text.

In Hargreaves's study, *Changing Teachers, Changing Times: Teachers' Work and Culture in a Postmodern Age* (1994), the interviews with 40 teachers and principals generated almost 1,000 pages of transcripts, which were read and reread in order to establish a close familiarity with the data. Summary reports of each interview were written according to the key themes. Themes appearing in the text were registered, classified, and reclassified on the basis of an active search for confirming and disconfirming evidence in the interviews. Hargreaves interrogated the data using a consciously eclectic approach, drawing in different concepts and theories. He describes his analytic approach as listening to the teacher's voices telling about their work and comparing their descriptions with claims about their work from literature. "Throughout the study, I have attempted to sustain a creative dialogue between different theories and the data, in a quest not to validate any presumed perspective, but simply to understand the problems in their social context, as experienced by teachers" (Hargreaves, 1994, p. 122).

No tables or quantified categorizations of themes are presented in the book. The findings are reported in a continuous interpretative text, with interview passages interspersed, such as the sequence leading to the concept of "contrived collegiality" (Box 1.2) that appeared unexpectedly from reading of the transcripts: "In contrived collegiality, collaboration among teachers was compulsory, not voluntary; bounded and fixed in time and space; implementation—rather than development—oriented; and meant to be predictable rather than unpredictable in its outcomes" (Hargreaves, 1994, p. 208). In the book, interview passages are integrated in theoretically informed reflections on teacher work, drawing on management literature and postmodern analysis of culture. The results are striking descriptions of the Canadian teachers' work situation in a postmodern culture, in particular regarding increasing pressures on time and collegiality, descriptions that are recognizable to teachers in Denmark.

Bourdieu and his colleagues offer few explicit textual interpretations of their interviews in *The Weight of the World* (1999). Although elsewhere he writes extensively and theoretically on the situation of the downtrodden in France, in this book Bourdieu mainly lets the many interviews reproduced in the book speak for themselves. However, the reader is aided in several ways:

It seems to me imperative to make explicit the intentions and procedural principles that we have put into practice in the research project whose findings we present here. The reader will thus be able to reproduce in the reading of the texts the work of both construction and understanding that produced them. (Bourdieu et al., 1999, p. 607)

When it is possible in the book to let the interviews to a large extent speak for themselves, this may be because a presentation of the social situation of the interviewees has paved the way, allowing the reader to interpret their statements in relation to their life situation. Further, much of the analysis was already built into the interviews through a Socratic maieutic method of aiding explanations, such as formulating suggestions for open-ended continuations of interviewee statements (see Chapter 8 on confronting interviews), Bourdieu also presents some of his own impressions from the interviews, such as the emotional impact upon himself of the interview with the two young men (Box 1.3):

> I did not have to force myself to share in the feeling, inscribed in every word, every sentence, and more especially in the tone of their voices, their facial expressions or body languages, of the *obviousness* of this form of collective bad luck that attaches itself, like a fate, to all those that have been put together in those sites of *social relegation*, where the personal suffering of each is augmented by all the suffering that comes from coexisting and living with so many suffering people together—and, perhaps more importantly, of the destiny effect from belonging to a stigmatized group. (Bourdieu et al., 1999, p. 64)

A theoretical reading of interview texts can draw in new contexts for regarding the interview themes, and bring forth new dimensions of familiar phenomena. A theoretical reading of texts may, however, also imply biased interpretations, with the readers only noticing those aspects of the phenomena that can be seen through their theoretical lenses. Notorious—and in our opinion often exaggerated—examples are Marxists only seeing the effects of class conflicts and psychoanalysts only seeing the manifestations of unconscious sexual forces in the phenomena they are studying. Theoretical bias is difficult to counteract; one measure, suggested when treating meaning interpretation in Chapter 12, is to make the analytic questions at the base of an interpretation explicit, and, in line with a hermeneneutic approach, to carefully reflect on one's own presuppositions concerning the research topic. Another option is to play the devil's advocate concerning one's own reading, trying to falsify it, and developing alternative interpretations, which may enable a distance from one's understanding, thereby counteracting a theoretical bias. That we understand the world in terms of our preunderstandings is not just an issue pertaining to what we here call theoretical reading, but seems to be a general feature of the human condition.

A theoretical reading may, however, in some instances block seeing new, previously not recognized, aspects of the phenomena being investigated. As one measure against making premature conclusions in interviewing, we emphasized in Chapter 7 careful listening, the importance of the interviewer being open to and sensitive toward the many nuances of what the interview subjects are telling them. For therapeutic interviews this listening attitude has been described as a freely floating attention. Also when reading the textual forms of the interviews the researcher may in one reading phase try to cultivate a sensitivity to the modes of expression and the multilayered meanings of the interview texts, attempting to adopt a phenomenological suspension, placing his or her own conceptions in brackets, in a attempt to attain a maximum openness to the texts as they present themselves.

Finally, the person of the reader attains a key importance with regard to interview analysis as theoretical reading; this concerns his or her personal sensitivity toward the subject matter as well as conceptual mastery of the theories applied in the reading of the interview texts. Here creativity is required in putting forth new interpretations and rigorousness in testing the interpretations. The theoretical interpretations of the interview texts are not validated by an adherence to a specific methodical procedure; the burden of proof remains with the researcher, on his or her ability to present the premises for, and to rigorously check, the interpretations put forth, and ability to argue convincingly for the credibility of the interpretations made. Engaging in what we call theoretical reading should not, however, be an excuse to proceed in a naive or loosely speculative manner. We return to this topic in the following chapters on validation and presentation of interview research.

Implications for Interviewing: For a theoretical analysis of interview texts, it is important that there is a rich material on those aspects of the subject matter relevant to the theoretical approaches (see Chapter 6 for the different contexts of teasing required for different types of theoretical interpretations). If the theoretical perspectives are not considered until the analysis stage, the interviews may lack the relevant information for making specific interpretations on the basis of a theory.

****

In the last three chapters, we have focused on different approaches to analyses that consider first the meaning, second the linguistic and discursive forms of interview texts, and third the analysis as bricolage and theoretical reading. With these approaches to analysis in mind, we once again direct our attention to interview research as a craft. The craftsman may have a large toolbox, but tools, just like social science research methods, do not dictate their own application, nor will the availability of tools as such lead to interesting results. The quality of the product is determined by the competence of the craftsman who applies the tools. A mechanical application of methods does not guarantee knowledge that is objective, reliable, valid, or generalizable. In the next chapter, we discuss these key social science concepts and seek to strip them of their metaphysical overtones and bring them down to earth, to the concrete practice and craftsmanship of qualitative research.

FIFTEEN

THE SOCIAL
CONSTRUCTION OF VALIDITY

———•◆•———

We now turn to the issue of how to get beyond the extremes of a subjective relativism in interview research, where everything can mean everything, and an absolutist quest for the one and only true, objective meaning. The trustworthiness, the strength, and the transferability of knowledge are in the social sciences commonly discussed in relation to the concepts of reliability, validity, and generalization. In this chapter we address the implications of these concepts for interview research. The main emphasis will be on validation—treating the interdependence of philosophical understandings of objectivity and truth, social science concepts of validity, and the practical issues of verifying interview knowledge.

In the first part of the chapter, we discuss principal issues of validation and attempt to draw the concepts of objectivity and validity back to everyday language and everyday activities. Thereafter we address the validation of interview knowledge in practice, where we argue that validation should not be confined to a separate stage of an interview inquiry, but rather permeate all stages from the first thematization to the final reporting. We then outline validation as pertaining to the quality of the craftsmanship of the researcher throughout an interview inquiry, and after that we depict communicative and pragmatic forms of validation. Finally, statistical and analytical generalization from interview studies is addressed.

OBJECTIVITY OF INTERVIEW KNOWLEDGE

Issues of reliability and validity go beyond technical or conceptual concerns and raise epistemological questions about the objectivity of knowledge and the nature of interview research. The question is whether knowledge produced through interviews can be objective. *Objectivity* is a rather ambiguous term, and we distinguish here between different meanings of objectivity relevant to qualitative research by taking our point of departure from everyday language and activities, such as when we speak about a journalistic interview as more or less objective. We will differentiate among the uses of objectivity: as freedom from bias, as reflexivity about presuppositions, as intersubjective consensus, as adequacy to the object, and as the object's ability to object.

Objectivity as *freedom from bias* refers to reliable knowledge, checked and controlled, undistorted by personal bias and prejudice. Such a common-sense conception of objective as being free of bias implies doing good, solid, craftsmanlike research, producing knowledge that has been systematically cross-checked and verified. In principle, a well-crafted interview can be an objective research method in the sense of being unbiased. This first meaning of objectivity is taken from ordinary language, where the demand to be objective is first and foremost an ethical demand ("You're not being objective about this!"), and secondarily an epistemological demand. Before we value the objectivity of scientific theories, we value the ethical person who sees other people as they are, and who does not impose his or her own biases on them. As Alasdair MacIntyre (1978) has argued: "Objectivity is a moral concept before it is a methodological concept, and the activities of natural science turn out to be a species of moral activity." (p. 37).

We may also speak of a *reflexive objectivity* in the sense of being reflexive about one's contributions as a researcher to the production of knowledge. Objectivity in qualitative inquiry here means striving for objectivity about subjectivity. In the language of hermeneutics, we can only make informed judgments, for example, in research reports, on the basis of our pre-judices (literally pre-judgments) that enable us to understand something (see Gadamer, 1975). The researcher should attempt to gain insight into these unavoidable prejudices and write about them whenever it seems called for in relation to the research project. Striving for sensitivity about one's prejudices, one's subjectivity, involves a reflexive objectivity.

A conception of objective as meaning *intersubjective* knowledge has been common in the social sciences. With objectivity as intersubjective consensus, we may distinguish between an arithmetical and a dialogical conception of objectivity. *Arithmetic intersubjectivity* refers to reliability as measured statistically by the degree of concurrence among independent observers or coders. Interview analyses may in principle be objective in the sense of intersubjective agreement, such as when a high degree of intersubjective reliability is documented by coding interviews in quantifiable categories. *Dialogical intersubjectivity* refers to agreement through a rational discourse and reciprocal criticism between those interpreting a phenomenon. This may take the form of a communicative validation among researchers as well as between researchers and their subjects. Taking into account the power asymmetry of researcher and subjects, the interview attains a privileged position regarding objectivity as dialogical subjectivity—the interview is a conversation and a negotiation of meaning between the researcher and his or her subjects.

Objective may also mean reflecting the nature of the object researched, letting the object speak, being *adequate to the object* investigated, an expression of fidelity to the phenomena, expressing the real nature of the object studied. Thus, if one conceives of the human world as basically existing in numbers, a restriction of the concepts of objectivity and validity to measurement follows naturally, as only quantitative methods then reflect the real nature of the social objects investigated. However, with the object of the interview immersed in a linguistically constituted and interpersonally negotiated social world, the qualitative research interview obtains a privileged position in producing objective knowledge of the social world. In Chapter 3, we sought to describe the object of interview knowledge according to seven characteristics, and the interview can be sensitive to and reflect the nature of the object investigated—a conversational human world. In the interview conversation, the object speaks.

Objectivity may also stand for *allowing the object to object*. Latour (2000) has suggested that it is by allowing the objects investigated to object to the natural scientists' interventions that maximum objectivity is obtained. This idea goes back to Dewey's pragmatism, where an object was simply defined as "that which objects, that to which frustration is due" (Dewey, 1925/1958, p. 239). Objectivity is attained when objects reveal themselves through acts that frustrate the researcher's preconceived ideas. In the social sciences, however, "nothing is more difficult than to find a way to render objects able to object to the utterances that we make about them" (Latour, 2000, p. 115). In contrast to the nonhuman objects of the natural sciences, human beings are incredibly complacent, "behaving all too easily as if they had been mastered by the scientist's aims and goals" (Latour, 1997, p. xv). What the human sciences should do, then, is not to look for cases where the human subjects are under optimal control, unable to influence the results of the research, or where the subjects are unaware of the real purpose of the investigation and "play the role of an idiotic object perfectly well" (Latour, 2000, p. 116) as in most psychological experiments, for example. Rather, in order to attain a degree of objectivity comparable to that which is possible in the natural sciences, the objects of a social study should be allowed to be "interested, active, disobedient, fully involved in what is said about themselves by others" (p. 116). If social scientists want to become objective, they should, as natural scientists do, seek the rare, extreme situations where their objects have maximum possibilities of protesting against what the researchers say about them; where the objects are allowed to raise questions in their own terms and not in the researcher's terms, a researcher whose interests they need not share. As an example from the social sciences Latour points to how feminism today has contributed to making women recalcitrant and likely to react against social researchers' interview approaches. We may here note that, in contrast to questionnaires, qualitative interviews, at least in principle, allow the subjects to object to the presuppositions of the researcher's questions and interpretations, an option that in particular is possible in what we have termed epistemic and confrontational interviewing.

We may conclude that, contrary to common opinion, knowledge produced in interviews need not be subjective, but qualitative interviews may, in principle, be an objective mode of inquiry with respect to several key meanings of objectivity. We now turn to the more specific discussions of reliability and validity and later generalization.

RELIABILITY AND VALIDITY OF INTERVIEW KNOWLEDGE

Some qualitative researchers have ignored or dismissed questions of validity, reliability, and generalization as stemming from oppressive positivist concepts that hamper a creative and emancipatory qualitative research. Other qualitative researchers—Lincoln and Guba (1985), for instance—have gone beyond the relativism of a rampant antipositivism and have reclaimed ordinary language terms to discuss the truth value of their findings, introducing concepts such as trustworthiness, credibility, dependability, and confirmability to qualitative research. From a feminist poststructural frame valorizing practice, Lather

(1995) addresses validity as an incitement to discourse, a fertile obsession, and attempts to reinscribe validity in ways that use the postmodern problematic to loosen the master code of positivism.

We retain the traditional concepts of reliability and validity, which are also terms in common language—as reflected in sentences such as "Your passport is not valid," "Your argument is not valid," and "Is he reliable?"—with the concept of reliability having not only a methodological but also a moral meaning, as when we speak of a reliable person. Here we will reinterpret these everyday concepts in ways appropriate to the production of knowledge in interviews. The present approach is thus not to reject the concepts of reliability, validity, and generalization, but to reconceptualize them in forms relevant to interview research. The understanding of verification starts in the lived world and daily language where issues of reliable craftsmen and reliable observations, of valid arguments, of transfer from one case to another, are part of everyday social interaction.

Reliability pertains to the consistency and trustworthiness of research findings; it is often treated in relation to the issue of whether a finding is reproducible at other times and by other researchers. This concerns whether the interview subjects will change their answers during an interview and whether they will give different replies to different interviewers. Issues of reliability during interviewing, transcribing, and analyzing have been treated in the previous chapters. Interviewer reliability was in particular discussed in relation to leading questions, which—when they are not a deliberate part of an interviewing technique—may inadvertently influence the answers, such as in the example of different wordings of a question about car speeds leading to different answers (Chapter 9). Interviewer reliability in the grade study was discussed on the basis of the categorizations of the subjects' answers (Chapter 12). In relation to transcription of interviews, an example was given of the intersubjective reliability of the transcripts when the same passage was typed by two different persons (Chapter 10). During categorization of the grading interviews, percentages were reported for the intersubjective agreement between two coders for the same interviews (Chapter 12). Although increasing the reliability of the interview findings is desirable in order to counteract haphazard subjectivity, a strong emphasis on reliability may counteract creative innovations and variability. These are more likely to follow when interviewers are allowed to follow their own interview styles and to improvise along the way, following up promising new hunches.

Validity refers in ordinary language to the truth, the correctness, and the strength of a statement. A valid inference is correctly derived from its premises. A valid argument is sound, well-grounded, justifiable, strong, and convincing. Validity has in the social sciences pertained to whether a method investigates what it purports to investigate. In a methodological positivist approach to social science, validity became restricted to measurement; for instance, "Validity is often defined by asking the question: Are you measuring what you think you are measuring?" (Kerlinger, 1979, p. 138). Qualitative research is then invalid if it does not result in measurements. In a broader conception, validity pertains to the degree that a method investigates what it is intended to investigate, to "the extent to which our observations indeed reflect the phenomena or variables of interest to us" (Pervin, 1984, p. 48). With this open conception of validity, qualitative research can, in principle, lead to valid scientific knowledge.

Many social science textbooks on methodology have been based on positivist epistemological assumptions, with a correspondence theory of truth. The standard definitions of validity have been taken from the criteria developed for psychological tests as formalized by Cronbach and Meehl (1955). In psychology, validity became linked to psychometrics, where the concurrent and predictive validity of the psychological tests were declared in correlation coefficients, indicating correspondence between test results and some external criteria. These psychometric tests, such as intelligence tests, have frequently been applied to predict school success. Cronbach and Meehl also introduced the concept of "construct validity," which originally pertained to measurements of theoretical constructs such as intelligence and authoritarianism; it involves correlations with other measures of the construct and logical analysis of their relationships. It was later extended by Cronbach to qualitative summaries as well as quantitative scores, where validation became an open process in which to validate is to investigate—"validation is more than corroboration; it is a process for developing sounder interpretations of observations" (Cronbach, 1971, p. 433). Later, Cherryholmes (1988) argued that construct validity is a discursive and rhetorical concept, open for phenomenological and deconstructive analysis.

The issue of what is valid knowledge involves the philosophical question of what is truth. Within philosophy, three classical criteria of truth are discerned—correspondence, coherence, and pragmatic utility. The *correspondence* criterion of truth concerns whether a knowledge statement corresponds to the objective world. The *coherence* criterion refers to the consistency and internal logic of a statement. And the *pragmatic* criterion relates the truth of a knowledge statement to its practical consequences. Although the three criteria of truth do not necessarily exclude each other, they have each obtained strong positions in different philosophical traditions. The correspondence criterion has been central within a positivist social science, where the validity of knowledge is expressed as its degree of correspondence with an objective reality. The coherence criterion has been strong in mathematics and hermeneutics. The pragmatic criterion has prevailed in pragmatism and to a certain extent in Marxist philosophy. The three truth criteria can be regarded as abstractions from a unity, where a comprehensive verification of qualitative research findings will involve observation, conversation, and interaction.

In a postmodern era, truth is constituted through a dialogue; valid knowledge claims emerge as conflicting interpretations and action possibilities are discussed and negotiated among the members of a community. Some implications of the above discussion for validation of qualitative research will now be taken up, leading to a conception of validity as quality of craftsmanship, as communication, and as pragmatic action.

First, when giving up a correspondence theory of truth as the basis for understanding validity, there is, following Popper, a change in emphasis from verification to falsification. The quest for absolute, certain knowledge is replaced by a conception of defensible knowledge claims. Validation becomes the issue of choosing among competing and falsifiable interpretations, of examining and providing arguments for the relative credibility of alternative knowledge claims (Polkinghorne, 1983). In an alternative concept of validity— going from correspondence with an objective reality to defensible knowledge claims—validity is ascertained by examining the sources of invalidity. The stronger the falsification attempts that a proposition has survived, the more valid and trustworthy the knowledge. Validation comes to depend on the quality of craftsmanship during an investigation, on continually checking, questioning, and theoretically interpreting the findings.

Second, a modern belief in knowledge as a mirror of reality recedes and a social construction of reality, with coherence and pragmatic criteria of truth, comes to the foreground. Method as a guarantee of truth dissolves; with a social construction of reality the emphasis is on the discourse of the community. *Communication* of knowledge becomes significant, with aesthetics and rhetoric entering into a scientific discourse.

Third, with a modern legitimation mania receding, there is an emphasis upon a *pragmatic* proof through action. The legitimation of knowledge through external justification, such as by appeals to some grand systems, or meta-narratives, and the modern fundamentalism of securing knowledge on some undoubtable, stable fundament, lose interest. Justification of knowledge is replaced by application; knowledge becomes the ability to perform effective actions. Criteria of efficiency and their desirability become pivotal, raising ethical issues of right action. Values do not belong to a realm separated from scientific knowledge, but permeate the creation and application of knowledge.

VALIDITY AS QUALITY OF CRAFTSMANSHIP

We will here attempt to demystify the concept of validity, to bring it back from philosophical abstractions to the everyday practice of scientific research. The concept of validity as quality of craftsmanship is not limited to a postmodern approach, but becomes pivotal with a postmodern dismissal of an objective reality against which knowledge is to be measured. The craftsmanship and credibility of the researcher becomes essential. Based on the quality of his or her past research in the area, the credibility of the researcher is an important aspect of fellow researchers ascribing validity to the findings reported. Validity is not only a matter of conceptualization and of the methods used; the person of the researcher (Salner, 1989), including his or her moral integrity (Smith, 1990) and especially what we called practical wisdom in our discussion of ethics, are critical for evaluating the quality of the scientific knowledge produced.

Although treated in a separate chapter here, validation does not belong to a separate stage of an investigation, but permeates the entire research process. We are here moving the emphasis from a final product validation to a continual process validation.

BOX 15.1   Validation at Seven Stages

1. Thematizing. The validity of an investigation rests upon the soundness of the theoretical presuppositions of a study and upon the logic of the derivations from theory to the research questions of the study.

2. Designing. The validity of the knowledge produced involves the adequacy of the design and the methods used for the subject matter and purpose of the study. From an ethical perspective a valid research design involves beneficience—producing knowledge beneficial to the human situation while minimizing harmful consequences.

3. Interviewing. Validity here pertains to the trustworthiness of the subject's reports and the quality of the interviewing, which should include a careful questioning to the meaning of what is said and a continual checking of the information obtained as a validation in situ.

4. Transcribing. The question of what is a valid translation from oral to written language is involved by the choice of linguistic style of the transcript.

5. Analyzing. This involves the question of whether the questions put to a text are valid and whether the logic of the interpretations made is sound.

6. Validating. This entails reflective judgment as to what forms of validation are relevant in a specific study and the application of the concrete procedures of validation, and a decision on what is the appropriate community for a dialogue on validity.

7. Reporting. This involves the question whether a given report gives a valid account of the main findings of a study, and also the question of the role of the readers of the report in validating the results.

Box 15.1 gives an overview of validity issues throughout an interview investigation. In the present approach, the emphasis on validation is not inspection at the end of the production line, but quality control throughout the stages of knowledge production. Are the steps in the research process each reasonable, defensible, and supportive of what the researcher concludes? Validation rests on the quality of the researcher's craftsmanship throughout an investigation, on continually checking, questioning, and theoretically interpreting the findings.

To Validate Is to Check. Validity is ascertained by examining the sources of invalidity. The stronger the falsification attempts a knowledge proposition has survived, the stronger and more valid is the knowledge. The researcher adopts a critical look upon the analysis, presents his or her perspective on the subject matter studied and the controls applied to counter selective perceptions and biased interpretations. The interviewer here plays the devil's advocate toward his or her own findings.

Various modes of checking findings have been suggested. An investigative concept of validation is inherent in the grounded theory approach of Glaser and

Strauss (1967). Validation is here not some final verification or product control; verification is built into the entire research process with continual checks on the credibility, plausibility, and trustworthiness of the findings. Kinsey and his colleagues (1948) went to great lengths to cross-check the validity of their subjects' reports, addressing factors such as memory and deliberate and unconscious cover-ups. During the interviews there was a continuous attention to the internal consistency of a subject's history, and there were later retakes of selected interviews and comparisons of pairs of histories from spouses.

Miles and Huberman (1994) emphasize that there are no canons or infallible decision-making rules for establishing the validity of qualitative research. Their approach is to analyze the many sources of potential biases that might invalidate qualitative observations and interpretations; they outline in detail tactics for testing and confirming qualitative findings. These tactics include checking for representativeness and for researcher effects, triangulating, weighing the evidence, checking the meaning of outliers, using extreme cases, following up on surprises, looking for negative evidence, making if-then tests, ruling out spurious relations, replicating a finding, checking out rival explanations, and getting feedback from informants (p. 263).

BOX 15.2   Why Did van Gogh Cut Off His Ear?

On the day before Christmas Eve in 1888, van Gogh cut off his left ear and gave it to a prostitute. More than a dozen explanations or interpretations of this act have been proposed in the psychobiographic literature. They range from inspiration from newspaper accounts of Jack the Ripper and influence from visits to bullfights in Arles, to aggression turned inwards, to appeals for sympathy, and to a reawakening of Oedipal themes.

Runyan (1981) has addressed these interpretations from the perspective of validating multiple interpretations in psychobiography. He asks if all of the interpretations are true, if some are true and others false, or if none of them are true. And do the various interpretations conflict, or do some of them supplement each other? From one point of view, the different interpretations provide a richly woven tapestry providing a fertile set of complementary explanations. A second way of making sense of the multiple interpretations would be that they are concerned with different aspects of a larger episode. A third approach works from the assumption that several of the interpretations may be valid while others are not. A fourth possible response is that the process of symbolic interpretation is hopelessly arbitrary.

Runyan adapts the third approach and critically evaluates the available evidence of the different interpretations. Concerning inspiration from Jack the Ripper, who mutilated the bodies of prostitutes, sometimes cutting of their ears, there is no direct empirical support that van Gogh had read the few stories in the local newspaper mentioning ear-cutting. In comparison, van Gogh's letters indicated that he had attended bullfights in Arles, which gives more support of influence from the bullfights, where the matador gives the ear of the bull to a lady of his choice. When it comes to the ear-cutting episode as an appeal for sympathy from his brother Theo, who was his main source of emotional and financial support, Runyan finds strong supportive evidence in the history of their relationship. Regarding the Oedipal interpretations, where van Gogh the day before the self-mutilation had used a razor to threaten Gauguin, with whom he was sharing a house and who was also a father figure to him, Runyan finds the biographical evidence for the razor episode rather meager.

After going through the different interpretations, attempting to identify faulty explanations and gathering corroborative evidence in support of others, Runyan suggests criteria for the evaluation interpretations in the light of (a) their logical soundness, (b) their comprehensiveness, (c) their survival tests of attempted falsification, (d) their consistency with the available evidence, (e) their support from more general knowledge of the person or the phenomenon, and (f) their credibility relative to other explanatory hypotheses.

Box 15.2 pictures, in relation to van Gogh's cutting off his ear, ways of approaching the validity of multiple interpretations, comparing the credibility and strength of different interpretations of an act. This includes checking the empirical evidence for and against an interpretation, and critically evaluating and comparing the relative plausibility of the different interpretations given for the same act. In general, the more attempts at falsification an interpretation has survived, the stronger it stands.

To Validate Is to Question. When ascertaining validity—that is whether an investigation investigates what it seeks to investigate—the content and purpose of the study precede questions of method. Different questions regarding "what" and "why" posed to interview texts lead to different answers of how to validate an interpretation. Discussing the question "Do photographs tell the truth?" Becker (1979) makes the general question "Is it true?" specific in "Is this photograph telling the truth about what?" And to decide what a picture is telling us the truth about, he suggests that we should ask ourselves what questions it might be answering. Concerning interview answers, an interviewee may not be "telling the truth" about factual states of affair, but the statements may still express the truth of the person's view of him- or herself, for example. Likewise, Hamlet's interview with Polonius may not uncover an objective truth about the shapes of clouds, but can rather lead us to valid knowledge about Polonius.

A common critique of research interviews is that their findings are not valid because the subjects' reports may be false. This is a possibility that needs to be checked in each specific case (see Dean & Whyte, 1969). The issue of validity depends on the "what" of the researcher's questions. In hermeneutical interpretations, the questions posed to a text become all important. Thus in the grading study, one type of research question led to an *experiential* reading of the pupils' statements about grades in high school, addressing the varieties of their individual experiences with grading. Another type of question led to a *veridical* reading, regarding the interviewees as witnesses or informants about the influence of grades on the social interaction in their class. The questioning also involved a *symptomatic* reading, focusing on the interviewees themselves and their individual reasons for making a given statement. Validation here differs with different questions posed to the interview texts (see Table 12.2).

To Validate Is to Theorize. Validity is not only an issue of method. Pursuing the methodological issues of validation generates theoretical questions about the nature of the phenomena investigated. In the terms of grounded theory, verifying interpretations is an intrinsic part of the generation of theory. Pursuing the methodological issues of validation generates theoretical and epistemological questions about the nature of the phenomena investigated. When treating discourse analysis in Chapter 13, we argued that if subjects frequently change their statements about their attitudes, for example to immigrants, during an interview, this is not necessarily due to an unreliable or invalid interview technique, but may in contrast testify to the sensitivity of the interview technique in capturing the multiple nuances and the fluidity of social attitudes.

The inconclusive results in the grade study of having informants triangulate and verify a pupil's belief in a connection between talkativeness and grades need not merely indicate a problem of method (Chapter 12); it also raises theoretical questions about the social construction of school reality. Pupils and teachers may live in different social realities with regard to which pupil behaviors lead to good grades. It is possible that pupils, in a kind of "superstitious" behavior, believe in a connection where there is none; or it may be that teachers overlook or deny a relation that actually exists. Ambiguity of the teacher's bases for grading, and contradictory beliefs by pupils and teachers about which behaviors lead to good grades, appear to be essential aspects of the social reality of school. The complexities of validating qualitative research need not be due to an inherent weakness in qualitative methods, but may on the contrary rest on their extraordinary power to picture and to question the complexity of the social reality investigated.

The quality of the craftsmanship in checking, questioning, and theorizing the interview findings leads ideally to transparent research procedures and convincing evident results. Appeals to external certification, or official validity stamps of approval, then become secondary, as validation is embedded in every stage of the production of knowledge throughout an interview inquiry.

COMMUNICATIVE VALIDITY

Communicative validity involves testing the validity of knowledge claims in a conversation. Valid knowledge is constituted when conflicting knowledge claims are argued in a conversation: What is a valid observation is decided through the argumentation of the participants in a discourse. In a hermeneutical approach to meaningful action as a text, Ricoeur (1971) rejected the position that all interpretations of a text are equal; the logic of validation allows us to move between the two limits of dogmatism and skepticism. Invoking the hermeneutical circle and criteria of falsifiability, he described validation as an argumentative discipline comparable to the juridical procedures of legal interpretation. Validation is based on a logic of uncertainty and of qualitative probability, where it is always possible to argue for or against an interpretation, to confront interpretations and to arbitrate between them.

A communicative approach to validity is found in several approaches in the social sciences. In psychoanalysis the validity of an interpretation is worked out in the interaction of patient and therapist. It is also implied in evaluation studies of social systems; House (1980) has thus emphasized that in system evaluation, research does not mainly concern predicting events, but rather whether the audience of a report can see new relations and answer new but relevant questions. Cronbach (1980) has advocated a discursive approach where validity rests on public discussion. The interpretation of a test is going to remain open and unsettled, the more so because of the role that values play in action based on tests; the aim for a research report is to advance sensible discussion—and, "The more we learn, and the franker we are with ourselves and our clientele, the more valid the use of tests will become" (p. 107). In a discussion of narrative research, Mishler (1990) has conceptualized validation as the social construction of knowledge. Valid knowledge claims are established in a discourse through which the results of a study come to be viewed as sufficiently trustworthy for other investigators to rely upon in their own work.

When conversation is the ultimate context within which knowledge is to be understood, as argued by Rorty (Chapter 3), the nature of the discourse becomes essential. There is a danger that a conception of truth as dialogue and communicative validation may become empty global and positive undifferentiated terms, without the necessary conceptual and theoretical differentiations worked out. Some specific questions concerning the how, why, and who of communication about validity will now be raised.

How. Communication can involve persuasion through rational discourse or through populist demagogy. The forms of persuasion about the truth of knowledge claims will be different in the harsh logical argumentation of a philosophical dialogue, in the juridical proceedings and legal interpretations in a courtroom, in a narrative capturing an audience, and in a humanistic therapy encounter based on positive feelings and reciprocal sympathy.

Philosophical discourses, such as the dialogues of Socrates, are characterized by a rational argumentation. The participants are obliged to test statements about the truth and falsity of propositions on the basis of argued points of view, and the best argument wins. This discourse is ideally a form of argumentation where no social exertion of power takes place, the only form of power being the force of the better argument.

Why. The question here concerns the purpose of a discourse about true knowledge. What are the aims and criteria of arriving at true knowledge? Habermas's (1971) normative discourse theory implies a consensual theory of truth (it thus differs from the descriptive discourse analysis in the tradition of Foucault): The ideal discourse aims at universally valid truths as an ideal. Eisner (1991) has advocated qualitative research as art, based on connoisseurship and criticism, accepting the personal, literary, and even poetic as valid sources of knowledge. The aim is here consensus: "Consensual validation is, at base, agreement among competent others that the description, interpretation, evaluation, and thematics of an educational situation are right" (Eisner, 1991, p. 112). From a postmodern perspective, Lyotard (1984) has, on the contrary, argued that consensus is only a stage in a discussion, and not its goal, which he posits as paralogy—to create new ideas, new differentiations, new rules for the discourse. To Lyotard, discourse is a game between adversaries rather than a dialogue between partners.

Who. Communicative validation implicates different participants in a community of validation. When the interviewer's interpretations refer to the subjects' own understanding of their statements, the interviewee becomes the relevant partner for a conversation about the correct interpretation, involving what has been termed "member validation" (see Table 12.2). The researcher's interpretations may also go beyond the subjects' self-understanding—what they themselves feel and think about a topic—while remaining within a critical commonsense understanding, such as in the case of the interpretation of the many denials of competition into a confirmation and with the deliberations of a jury on the trustworthiness of a witness. The general lay public is the relevant community of this "audience validation." In a third context, a theoretical frame for interpreting the meaning of a statement is applied. The interpretations are then likely to go beyond the interviewees' self-understanding and also to exceed a commonsense understanding, for example in an interpretation of grades as the currency of the educational system, with a built-in contradiction of the use value and the exchange value of knowledge. In such cases the relevant community of validation consists of scholars familiar with the interview themes and with the theories applied to the interview texts, and this can be referred to as "peer validation." In the present perspective, none of the three contexts provide more correct, authentic, or valid knowledge than the others. They are each appropriate to different research questions, which may be posed to interview statements.

Validation through negotiations of the community of scholars is nothing new; in the natural sciences the acceptance of the scientific community has been the last, ultimate criterion for ascertaining the truth of a proposition. What is relatively new in qualitative research in the social sciences is the emphasis on truth as negotiated in a local context, with extension of the interpretative community to include the subjects investigated and the lay public.

Communicative validation approximates an educational endeavor where truth is developed in a communicative process, with both researcher and subjects learning and changing through the dialogue, a process which may also involve the larger public, such as in Bellah's conception of social science as "public philosophy" (Bellah et al., 1985).

A heavy reliance on intersubjective validation may, however, also imply a lack of work on the part of the researcher and a lack of confidence in his or her interpretations, with an unwillingness to take responsibility for the interpretations. There may be a general populist trend when leaving the validation of interpretations to the readers, as in reader response validation, with an abdication to the ideology of a consumer society: "The customer is always right."

Power and Truth. Different professional communities may construct knowledge differently, and conflicts may arise about which professions have the right to decide what is valid knowledge within a field, such as health, for example. Furthermore, there is the specific issue of who decides who is a competent and legitimate member of the interpretative community. The selection of members of the community to make decisions about issues of truth and value is considered crucial for the results in many cases, such as in the selection of members of a jury, or of a committee to examine a doctoral candidate, or of an academic appointment committee.

Habermas's consensus theory of truth is based on the ideal of a dominance-free dialogue, which is a deliberate abstraction from the webs of power relationships within real-life discourses, in contrast with Lyotard's postmodern understanding of a scientific conversation as a game of power. More generally, Lyotard argues, scientists are not purchased to find truth, but to augment power: "The games of scientific language become the games of the rich, in which whoever is wealthiest has the best chance of being right. An equation between wealth, efficiency and truth is thus established" (Lyotard, 1984, p. 45).

PRAGMATIC VALIDITY

Pragmatic validation is verification in the literal sense—"to make true." To pragmatists, truth is whatever assists us to take actions that produce the desired results. Knowledge is action rather than observation, the effectiveness of our knowledge beliefs is demonstrated by the effectiveness of our action. In the pragmatic validation of a knowledge claim, justification is replaced by application. Marx (1888/1998) stated in his second thesis on Feuerbach that the question of whether human thought can lead to objective truth is not a theoretical but a practical one. Humans must prove the truth, that is, the reality and power of their thinking in practice. And his 11th thesis is more pointed; the philosophers have only interpreted the world differently, what matters is changing the world.

From a hermeneutical angle, Taylor (1985) also emphasizes the pragmatic aspects of validity in the social sciences. He argues that social theory is itself a kind of practice that serves to interpret and articulate the meanings of human activity, but these articulations may again enter the actors' self-understandings, thereby changing the realities they are concerned with. Taylor's point is that validity in the social sciences cannot mean mirroring some independent objects researched, for the objects of human and social science are not independent of, but rather constituted by, human understanding. Instead, he argues, validity in the social sciences means *improving* the practices under consideration, which is as much a moral and political issue as it is a question of epistemology. New qualitative interpretations can alter the self-understandings of those they describe, and the validity of social theories can thus be tested by examining the quality of the practices they inform and encourage. More recently, Flyvbjerg (2001) has relied on Aristotle's notion of *phronesis* (which we discussed in Chapter 4) to develop a framework for what he calls "phronetic social science," getting close to the practices studied, focusing on values, employing case studies and narrative, with the overall goal of producing input "to the ongoing social dialogue and praxis in a society, rather than to generate ultimate, unequivocally verified knowledge." (p. 139). A valid qualitative account would, from this pragmatic "phronetic" perspective, be one that contributed fruitfully to the public discussion about values and goals in a society, and Flyvbjerg mentions the study by Bellah and colleagues (1985) as a significant example of "phronetic social science."

A pragmatic concept of validity goes farther than communication; it represents a stronger knowledge claim than an agreement through a dialogue. Pragmatic validation rests on observations and interpretations, with a commitment to act on the interpretations: "Actions speak louder than words." With the emphasis on instigating change, a pragmatic knowledge interest may counteract a tendency of social constructionism and postmodernism to circle around in endless interpretations and deconstructions.

We may distinguish between two types of pragmatic validation—whether a knowledge statement is accompanied by action and whether it instigates changes of action. In the first case, validation of a subject's verbal statement is based on supporting action that *accompanies* the statement. This concerns going beyond mere lip service to a belief, to following it up with action. Thus in investigations of racial prejudice, comprehensive inquiries go beyond a subject's mere verbal statements against racial segregation and investigate whether the statements are also accompanied by appropriate supportive actions.

The second, stronger form of pragmatic validation concerns whether interventions based on the researcher's knowledge may *instigate* actual changes in behavior. In collaborative action research, investigators and subjects together develop knowledge of a social situation and then apply this knowledge through new actions in the situation, thus testing the validity of the knowledge in praxis. Reason (1994) describes a study of health workers that was based on participatory inquiry with a systematic testing of theory in live-action contexts. The topic was stress that came from hidden agendas in their work situation, such as suspicions of drug taking and of child abuse in the families the health workers visited. The co-researchers first developed knowledge through discussions among themselves, by role-playing, and thereafter by raising their concerns directly with their client families. Reason discusses the validity in this cooperative inquiry, and emphasizes the need to get beyond a mere consensus collusion where the researchers might band together as a group in defense of their anxieties, which may be overcome by a continual interaction between action and reflection throughout the participatory inquiry.

Freud did not rely on the patient's self-understanding and verbal communication to validate therapeutic interpretations; he regarded neither the patient's "yes" nor "no" to his interpretations as sufficient confirmation or disconfirmation; the "yes" or "no" could be the result of suggestion as well as of resistance in the therapeutic process. Freud recommended more indirect forms of validation, such as observing the patient's reactions after an interpretation, for example in the form of changes in the patient's free associations, dreams, recall of forgotten memories, and alteration of neurotic symptoms (Freud, 1963, p. 279). Spence (1982) has followed up on the emphasis on the pragmatic effects of interpretations: Narrative truth is constructed in the therapeutic encounter, it carries the conviction of a good story, and it is to be judged by its aesthetic value and by the curative effect of its rhetorical force. In therapeutic work different interpretations of a behavior can be tactically employed in a flexible approach to alter the behavior in question, such as in Scheflen's discussions of Susan's smile in Box 12.1.

How. The form of pragmatic validation can vary: It can be a patient's behavioral reactions to the therapists's interpretation of his or her dreams, the reactions of an audience to a system evaluation report, or the cooperative interaction of researcher and subjects in action research.

Why. A scientific discourse is, in principle, indefinite; there is no requirement of immediate action; new arguments that could alter or invalidate earlier knowledge can always appear. In contrast to the uncoerced consensus of the scientific discourse, practical contexts may require actions to be undertaken and decisions to be made that involve a coercion to consensus. This includes the proceedings of a jury, the negotiations of a dissertation committee, decisions about therapeutic interventions, and decisions about institutional changes in action research.

A pragmatic approach implies that truth is whatever assists us to take actions that produce the desired results. Deciding what the desired results are involves values and ethics. The importance of values in validation follows through a change of emphasis in social research from primarily mapping the social world with respect to *what is* to changing the focus to *what could be.* Thus Gergen's (1992) postmodern conception of generative theory involves research that opens new possibilities of thought and action as a means of transforming culture.

Who. The question of "who" involves the researcher and the users of the knowledge produced. Patton (1980) emphasizes the credibility of the researcher as an important criterion of whether a research report is accepted or not as a basis for action. Here the quality of the research craftsmanship and the reliability of the researcher as a person become critical to judgments of the validty of the research reported. The question of "who" also involves ethical and political issues. Who is to decide the direction of change? There may be personal resistance to change in a therapy as well as conflicting vested interests in the outcome of a system evaluation or an action study. Thus, regarding audience validation in system evaluation, who are the stakeholders that will be included in the decisive audience—the funding agency, the leaders of the system evaluated, the employees, or the clients of the system?

Power and Truth. Pragmatic validation raises the issue of power and truth in social research: Where is the power to decide what the desired results of a study will be, or the direction of change; what values are to constitute the basis for action? And, more generally, where is the power to decide what kinds of truth seeking are to be pursued, what research questions are worth funding? Following Foucault we should here beware of localizing power to specific persons and their intentions, and instead analyze the netlike organization and multiple fields of power-knowledge dynamics.

VALIDITY OF THE VALIDITY QUESTION

We have above argued for integrating validation into the craftsmanship of research, and for extending the concept of validation from observation to also include communication about, and pragmatic effects of, knowledge claims. We have further attempted to demystify the concept of validity, maintaining that verification of information and interpretations is a normal activity in the interactions of daily life. Even so, a pervasive attention to validation can be counterproductive and perhaps lead to a general invalidation. Rather than let the product, the knowledge claim, speak for itself, validation can here involve a legitimation mania that may further a corrosion of validity—the more one validates, the greater the need for further validation. Such a counter-factuality of strong and repeated emphasis on the truth of a statement may be expressed in the folk saying, "Beware when they swear they are telling the truth."

Ideally, the quality of the craftsmanship results in products with knowledge claims that are so powerful and convincing in their own right that they, so to say, carry the validation with them, like a strong piece of art. In such cases, the research procedures would be transparent and the results evident, and the conclusions of a study intrinsically convincing as true, beautiful, and good. Valid research would in this sense be research that makes questions of validity superfluous.

GENERALIZING FROM INTERVIEW STUDIES

If the findings of an interview study are judged to be reasonably reliable and valid, the question remains whether the results are primarily of local interest, or whether they may be transferable to other subjects and situations. A persistent question posed to interview studies is whether the results are generalizable. In everyday life we generalize more or less spontaneously. From our experience with one situation or person we anticipate new instances, we form expectations of what will happen in other similar situations or with similar persons. Scientific knowledge also lays claim to generalizability; in methodological positivist versions, the aim of social science was to produce laws of human behavior that could be generalized universally. A contrasting humanistic view implies that every situation is unique, each phenomenon has its own intrinsic structure and logic. Within psychology, natural science-oriented schools, such as behaviorism, have sought universal laws of behavior, whereas the uniqueness of the individual person has dominated in humanistic psychology. In a postmodern approach the quest for universal knowledge, as well as the cult of the individually unique, is replaced by an emphasis on the heterogeneity and contextuality of knowledge, with a shift from generalization to contextualization.

A common objection to interview research is that there are too few subjects for the findings to be generalized. A first reply to this objection is "Why generalize?" Consistent demands for the social sciences to produce generalizable knowledge may involve an assumption of scientific knowledge as necessarily universal and valid for all places and times, for all humankind from eternity to eternity. In contrast, pragmatist, constructionist, and discursive approaches conceive of social knowledge as socially and historically contextualized modes of understanding and acting in the social world.

Forms of Generalizing. In qualitative research, generalization has often been treated in relation to case studies. When and how can one generalize from a single case to other cases? Stake (2005) makes a distinction between three kinds of case studies: the intrinsic case study (undertaken in order to better understand this particular case), the instrumental case study (undertaken to provide insight into more general issues), and the multiple or collective case study (which is an instrumental case study extended to several cases). In contrast to traditional scientific demands for generalizability, Stake argues that intrinsic case studies are worthwhile in their own right, which is commonplace in the humanities, where one rarely questions the value of better understanding the work of an author, for example, simply for the sake of understanding itself.

If we are interested in generalizing, however, we may ask not whether interview findings can be generalized globally, but whether the knowledge produced in a specific interview situation may be transferred to other relevant situations. The second reply to the above question is then "How can one generalize?" Three forms of generalizing will be outlined below based on Stake's discussion of generalization from case studies—naturalistic, statistical, and analytic.

Naturalistic generalization rests on personal experience: It develops for the person as a function of experience; it derives from tacit knowledge of how things are and leads to expectations rather than formal predictions; it may become verbalized, thus passing from tacit knowing to explicit propositional knowledge.

Statistical generalization is formal and explicit: It is based on representative subjects selected at random from a population. Statistical generalization is feasible for interview studies using even a small number of subjects in so far as they are selected by random and the findings quantified (Chapter 6). With the use of inferential statistics the confidence level of generalizing from the selected sample to the population at large can be stated in probability coefficients. When the interviewees are selected at random and the interview findings quantified, the findings may be subjected to statistical generalization. Thus for the correlation found between talkativeness and grade point average it was possible to state that there was only 1 in 1,000 probability that this was a chance finding limited to the 30 randomly chosen pupils of the grade study (Chapter 12). For the application of powerful statistical tests, however, larger samples of subjects are required than are feasible for most interview studies.

Further, due to the involved statistical presuppositions, the findings of a self-selected sample, such as volunteers to a treatment, cannot be transferred to the population at large. In many cases interview subjects are selected by criteria such as typicality or extremeness, or simply by accessibility. For example, an interview sample of women who have turned to a help center for victims of violence is a self-selected and not a random sample from the population. Their strong motivation for help may lead to valuable knowledge on the nature of being subjected to violence. The findings of the self-selected sample cannot, however, be statistically generalized to the population at large.

Analytical generalization involves a reasoned judgment about the extent to which the findings of one study can be used as a guide to what might occur in another situation. It is based on an analysis of the similarities and differences of the two situations. In contrast to spontaneous naturalistic generalization, the researcher here bases the generalization claims on an assertional logic. There are several forms of assertational logic, such as the legal form of argumentation in court and arguments for generalization based on theory. By specifying the supporting evidence and making the arguments explicit, the researcher can allow readers to judge the soundness of the generalization claim (see also Yin, 1994, on inductive generalization). We may here discern a researcher-based and a reader-based analytical generalization from interview studies. In the first case the researcher, in addition to rich specific descriptions, also offers arguments about the generality of his or her findings. In the latter case it is the reader who, on the basis of detailed contextual descriptions of an interview study, judges whether the findings may be generalized to a new situation.

In a discussion of the use of case studies in system evaluation, Kennedy (1979) argues for establishing rules for drawing inferences about the generality of qualitative findings from a case study, rules of inference that reasonable people can agree on. As models for inspiration, she turns to generalization in legal and clinical practice. In case law it is the most analogous preceding case, the one with the most attributes similar to the actual case, that is selected as the most relevant precedent. The validity of the generalization hinges on an analysis of the similarities and differences between the original and the present case, on the extent to which the attributes compared are relevant, which again presupposes rich, dense, and detailed descriptions of the cases, or what we referred to as "thick description" in Chapter 4. Kennedy suggests criteria for relevant attributes of comparison in legal and clinical cases, which in the clinical situation are the precision of descriptions, longitudinal information, and multidisciplinary assessment. It is the receiver of the information who determines the applicability of a finding to a new situation. In case law, the court decides whether a previous case offers a precedent that can be generalized to the case being tried.

> Like generalisations in law, clinical generalisations are the responsibility of the receiver of information rather than the original generator of information, and the evaluator must be careful to provide sufficient information to make such generalisations possible. (Kennedy, 1979, p. 672)

The use of case studies in social research has often been criticized by traditional researchers. Flyvbjerg has addressed this critique, which he to a large extent attributes to misunderstandings about case studies.

BOX 15.3   Misunderstandings About Case Studies

Flyvbjerg (2006) rejects five misunderstandings about case study research:

1. *That general, context-independent knowledge is more important than concrete, context-dependent knowledge.* It can here be argued that there simply cannot be found universals in the study of human affairs, since human activity is situated in local contexts of practice, so, because of the nature of the human world, context-dependent knowledge is more valuable than a vain search for universal, predictive theory.

2. *That one cannot generalize on the basis of an individual case.* Against this misunderstanding, Flyvbjerg mentions a number of cases, where it has no doubt been possible to generalize from single case studies such as Galileo's repudiation of Aristotle's law of gravity, which was based on a single practical experiment with falling bodies (Galileo demonstrated that objects of different weights will fall at the same speed). Generalizability of case studies can be increased by the strategic selection of cases, for example by choosing a "critical case," one can approach a research design that permits logical deductions of the type "If this is (not) valid for this case, then it applies to all (no) cases" (Flyvbjerg, 2006, p. 230). A case study may thus, for example, function as a "black swan," falsifying generally accepted beliefs reached by scientific induction (e.g., that "all swans are white").

3. *That case studies are only useful for generating hypotheses.* Following from the rejection of the second misunderstanding, we here have an argument that case studies can be used to test and falsify hypotheses, rather than simply generate hypotheses. By considering extreme or critical cases, for example, researchers can sometimes falsify taken-for-granted assumptions about human conduct.

4. *That case studies contain a bias toward verification.* Against this fourth common misunderstanding, Flyvbjerg cites a number of researchers (e.g., Campbell, Ragin, Geertz) who report that their preconceived views were proven wrong by the case study, rather than automatically verified. It seems that if there is a bias in case study research, it is toward falsification rather than verification.

5. *That it is difficult to develop general theories on the basis of specific case studies.* In contrast to this final misunderstanding, Flyvbjerg argues that small-N qualitative research is often at the forefront of theoretical

development. Here it seems quite easy to come up with examples from most disciplines that support Flyvbjerg's intuition—in the case of psychology's classics we can mention qualitative researchers such as Freud, whose case study based theories are still debated, and in current educational research, some of the most influential work, which has given rise to significant developments in learning theory, has been based on case studies of apprenticeship (Lave & Wenger, 1991).

In Box 15.3 we have summarized and extended Flyvbjerg's arguments against some common misunderstandings about case studies. That a large number of cases or instances studied is not always necessary is well known in the natural sciences, where Harvey's discovery that the heart is a pump could in principle be arrived at through careful analysis of a single heart. Single case studies based on interviews can also be valuable if done well, and some famous examples can be found in Freud's writings, for example his analyses of Anna O., the Wolf man, and the Rat man.

Analytical generalization may be drawn from an interview investigation regardless of sampling and mode of analysis. Analytical generalization rests upon rich contextual descriptions and includes the researcher's argumentation for the transferability of the interview findings to other subjects and situations, as well as the readers' generalizations from a report. Analytical generalization and communicative validation both presuppose high-quality descriptions of the interview process and products. This points to the importance of how interview studies are reported, which we shall treat in the following chapter.

SIXTEEN

REPORTING
INTERVIEW KNOWLEDGE

R eporting research interviews does not simply mean re-presenting the views of the interviewees, accompanied by the researcher's viewpoints in the form of interpretations. The interview report is itself a social construction in which the author's choice of writing style and literary devices provide a specific view on the subjects' lived world. The writing process is one aspect of the social construction of the knowledge gained from the interviews, and the quality of an interview report attains a key position when validation and generalization of interview findings include communication with readers.

In this chapter, we first address some issues of reporting interviews; they concern reporting to different audiences, the frequently boring character of interview reports, and the ethics of reporting. Then we turn to the practical issues of writing an interview report, and we argue for integrating writing in the entire research process. We present in some detail standard modes of reporting interviews, and ways of improving these, such as the use of interview quotes. We conclude the chapter by giving examples of a variety of modes of enriching interview reports, ranging from dialogues to collages.

CONTRASTING AUDIENCES FOR INTERVIEW REPORTS

Throughout this book, we have treated interviews and interview research as forms of conversation. The interview researcher has been depicted as a traveler

in a foreign country, learning through his or her conversations with the inhabitants. When the traveler returns home with tales about encounters in the distant land, he or she may discover that local listeners react rather differently to
these stories. At times it may be easier for interview researchers to carry out
the conversations with their respondents rather than to enter into conversations
with colleagues about their interview conversations.

When writing a report from an interview study it may be useful to be aware
of different requirements within local social science communities. Authoritative
requirements for a social science report may differ markedly across departments, disciplines, countries, and epochs. In the previous discussion of interview quality some mainstream objections to the objectivity of qualitative
interview research were addressed (Box 9.3). Such objections are in line with a
modernist conception of interviewing as unearthing preexisting nuggets of data
and meanings from the depths of interview subjects. A contrasting audience
consists of researchers working within the qualitative field and the humanities,
who may come close to a postmodern understanding of the interviewer as a
traveler returning home with tales from conversations in a distant country.
Interviews are, as Holstein and Gubrium (1995) have argued, unavoidably
interpretively active, meaning-making practices. Researchers of a postmodern
sensibility are particularly sensitive to practices of representation, such as how
the lives of those interviewed are depicted in interview reports: "Today we
understand that we write culture, and that writing is not an innocent practice.
We know the world only through our representations of it." (Denzin, 2001,
p. 23). We write culture, and, as interview researchers, we also jointly speak and
write subjectivities into being when we interview and report our findings.

The closeness of interview studies to ordinary life, with their often lively
descriptions and engaging narratives, makes an interview report potentially
interesting to the general public. In some cases, this may entail a conflict
between the demands of the scientific and the general communities, between
presenting the results in a scientifically documented and controllable form or in
an illustrative and engaging popular form. The dilemma of presenting captivating stories versus formal documentation of method and findings may be envisaged by two contrasting scenes for the report—the art gallery and the courtroom.
An interview report should ideally be able to live up to artistic demands of
expression as well as to the rigor of the cross-examination of the courtroom.

In *art* it is the end product—a painting or a sculpture—that is essential,
and not the methods of the production process. The painting techniques employed

may be of interest to fellow artists and to art historians, but the techniques are not the reason for taking a piece of art seriously. A painting carries its own message, it convinces through its expression and style. In literature, the content and form of Shakespeare's dramas still capture us today, while little is known about the dramas' origins or of Shakespeare's methods of writing.

In contrast, in a report to a *court*, say from interviews by a psychologist about child abuse, eloquence and style are not essential to the report. There will be an intense cross-examination from the prosecution and the defense, trying to find weak points in the interviews and their interpretations. What is crucial in a courtroom is not just what lawyers or witnesses believe, but whether they are able to justify their beliefs with good reasons when challenged. This gives conversations in the courtroom an *epistemic* dimension, as addressed in Chapter 3 (*episteme* is knowledge that has been justified discursively in a conversation). The procedures will be under scrutiny and attempts made to undermine the reliability of witnesses; of the forms of interrogation, such as the influence of leading questions; and the logic of the interpretations drawn.

BORING INTERVIEW REPORTS

Some 3,000 years ago, Odysseus returned to Greece from his research inquiry in distant countries. Homer's oral tale of the voyage, later written down, was cast in a form that still engages today. Freud's therapeutic case stories from a century ago still provoke heated controversies. Current interview studies may not be that long lived; reports need to be read to have a life after publication. Some impressions from reading current interview reports will be offered.

Tiresome Interview Findings. Interview studies are sometimes tedious to read: They are frequently characterized by long, obtuse, verbatim quotes, presented in a fragmented way, with primitive categorizations, and not seldom at inflated length. Hundreds of pages with quotes from the interview transcripts, interspersed with some comments and a few tables with numbers from categorizations, seldom make for interesting reading. The subjects' often exciting stories have—through the analyzing and reporting stages—been butchered into atomistic quotes and isolated variables. This style of reporting interviews may have been influenced by a qualitative hyperempiricism, with the many interview quotes made to serve as basic facts. Extensive verbatim transcripts are regarded as rock-bottom documentation of what was really said in the interviews. The different rhetorical forms of oral and written language are overlooked in the construction of verbatim interview transcripts, with their tiresome repetitions, fillers, and incomplete sentences.

After having endured the reading of a series of interview reports, one may long for some sociological and anthropological case studies, as well as dramatic narrative therapeutic case histories, that can both be entertaining and carry provocative new insights. One may even look forward to reading about laboratory experiments with their neat logical rigor, elegant designs, clear presentations, and stringent discussions of the findings and considerations of possible sources of error that could invalidate the findings.

Readers, who find this description unduly negative, are referred to Richardson's more devastating criticism of interview reports: "I confessed that for years I had yawned my way through numerously exemplary of qualitative studies. Countless numbers of texts had I abandoned half read, half scanned. I would order a new book with great anticipation . . . only to find the text boring" (Richardson & Adams St. Pierre, 2005, p. 959).

There may be several reasons for colorless interview reports. The writer may be so overwhelmed by the extensive and complex interview texts that any personal perspective on the interviews is lost. The researcher may strongly identify with the interview subjects, "go native," and be unable to retain a conceptual and critical distance from the subjects' accounts. The fear of subjective interpretations may lead to reports that consist of a wearisome series of uninterpreted quotes, with the researcher refraining from theoretical interpretations as if from some dangerous form of speculation. The page inflation of interview reports may simply be due to researchers not knowing what story they want to tell, and they therefore are not able to select the main points they want to get across to their audience. Without knowing the "what" and the "why" of the story, the "how"—the form of the story—becomes problematic.

Method as a Black Box. If readers actually find the interview results of interest, they may want to know about the design and the methods that have produced this intriguing knowledge. They are then likely to encounter a black box. The readers will have to guess about the social context of the interview, the instructions given to the interviewees, the questions posed, and the procedures used during transcribing and analyzing the interviews. For a reader who wants to evaluate the trustworthiness of the findings, to reinterpret or apply the results, information on the methodic steps of an investigation is mandatory. In interview reports, however, the link between the original conversations and the final report is often missing.

Qualitative interviews can contain detailed descriptions of the subjects' life situations, their experiences and actions, but may be virtually devoid of descriptions of the interview situation and of the researcher's actions used to obtain the information reported about the subjects. Though the strengths of qualitative studies are their detailed descriptions and the use of the researcher as an instrument, depictions of the researcher's own activities while producing the knowledge are often conspicuously absent.

One reason for the neglect of method may be that an interview study hardly follows discrete, formal procedures; much is left to improvisation and the intuition of the interviewer and interpreter. A further reason may be that there are no established common conventions for reporting qualitative studies. Rather than leading to a silence on method, the unique nature of an interview study should in fact pose a challenge to the researcher to describe as precisely as possible the specific steps, procedures, and decisions taken in the specific study. Everything is potentially important to our understanding of the process of knowledge production—and could potentially be included in the final report. It is a virtue in qualitative research to describe these elements, whenever they are relevant. This demand for methodological transparency can be taken to the extreme, as testifies the advice given by Latour, who believes that we as researchers should

> keep track of all our moves, even those that deal with the very production of the account [i.e. the report]. This is neither for the sake of epistemic reflexivity nor for some narcissist indulgence into one's own work, but because from now on *everything is data:* everything from the first telephone call to a prospective interviewee, the first appointment with the advisor, the first corrections made by a client on a grant proposal. . . ." (Latour, 2005, pp. 133–134)

If everything is data, interview researchers should develop practical ways of keeping track of what they are doing, which may involve logbooks and diaries of different sorts, and here they may learn from anthropologists doing fieldwork, who often work with a number of different books in which they register what they observe and learn, and also note personal reflections that may prove to be important and useful when reporting.

Before we turn to possible ways to improve interview reports with regard to scientific criteria of rigor and artistic criteria of elegance, some of the moral issues involved in publishing interviews will be addressed.

ETHICS OF REPORTING

The publication of a research report raises moral questions about what kinds of effects a report leads to. Richardson (1990) emphasizes that writing itself creates value, language is not simply reflecting a reality out there, but creates a particular view of reality. The author is a narrator, a person who speaks on behalf of others:

> Because writing is always value constituting, there are always the problems of authority and authorship. . . . Narrative explanations, in practice, mean that one person's voice—the writer's—speaks for that of the others. . . . These practices, of course, raise postmodernist issues about the researcher's authority and privilege. For whom do we speak and to whom do we speak, with what voice, to what end, using what criteria? (Richardson, 1990, pp. 26–27)

Bourdieu also points to the ethical responsibility of the author of an interview report, as "the analyst not only has to accept the role of transmitter of their [. . .] symbolic efficacy, but, above all, risks allowing people free play in the game for reading, that is, in the spontaneous (even wild) constructions each reader necessarily puts on things read (Bourdieu et al., 1999, p. 623).

Here we will discuss ethical issues when reporting interviews in relation to the ethical guidelines of informed consent, confidentiality, and consequences, as outlined in Chapter 4.

Informed Consent. As discussed earlier, care should be taken before the interview situation to have a clear understanding with the interviewees about the later use and possible publication of their interviews, preferably with a written agreement (Chapters 6 and 7).

Confidentiality. In order to protect the subjects' privacy, fictitious names and sometimes changes in subjects' characteristics are used in the published results. This requires altering the form of the information without making major changes of meaning. Yet disguising subjects is not without hazards. A misleading camouflage can be illustrated by an interview study of refugees' adaptation to the Danish culture. A student had in her master's thesis changed the names as well as the nationalities of the refugees she had interviewed and quoted at length, at the suggestion of her advisor, this book's first author. The external examiner pointed out a serious lack of understanding in the thesis's analysis of the social and psychological situation of a refugee from Chile. On closer examination, it turned out that the "Chilean" refugee was a disguised

Polish refugee. The student, herself an immigrant, had not taken into account that Polish refugees in Denmark in the 1970s tended to be strongly anticommunist and Chilean refugees to be equally strongly socialist or communist. Disguising names and nationality had brought about marked changes in the meaning of the social situations and identity of the subjects, whereby several of the interpretations made little sense. The example points out the problem of concealing information without substantially changing its meaning, which requires an extensive knowledge of the phenomena investigated.

The particular problems of privacy in the writing stage of a qualitative inquiry have been discussed by Glesne and Peshkin (1992), who mention several well-known social science studies in which reporters and others have been able to track down the actual persons, despite the use of fictitious names and the like. One of the more easily resolved issues of confidentiality involves interviewees who do not want to be anonymous subjects: They have become engaged in a project and want to take responsibility for their statements by having their full names on them.

Consequences. Qualitative research should ideally both produce scientific knowledge and contribute to ameliorating the human condition (see Chapter 4). This involves communicating the findings in a form that is both scientifically and ethically sound. It may be difficult for a researcher to anticipate the potential ethical and political consequences of an interview report. One unintended consequence of the grade study will be mentioned. A teacher of French, who had received a copy of the chapter containing the results from his interview, called and asked the first author not to use his statements in the book. In high school, French was an unpopular subject for many pupils and this teacher was keenly aware of and eloquent about his use of grades to motivate his pupils to learn French. His statements were highly illustrative of the use of grades as a motivational device and would be easily grasped by the readers. At the time of publication, however, a public discussion had started about the relevance of keeping French as a subject in Danish high schools. The teacher now feared that his descriptions of using grades to motivate his pupils to learn the unpopular language could be used in the public debate as an argument for omitting French as a school subject. The negative consequences did not directly concern the teacher himself, but rather his profession with regard to the public image of French as a school subject. His request was granted and "French" was changed to "English" in his statements, which thereby lost some of their expressive value.

Other decisions about whether to change a report due to anticipated consequences may not be so easily solved. Glesne and Peshkin (1992) raise a general question:

> What obligations does the researcher have to research participants when publishing findings? If the researcher's analysis is different from that of participants, should one, both, or neither, be published? Even if respondents tend to agree that some aspect of their community is unflattering, should the researcher make this information public? (p. 119)

The intended purpose of the grade study was to document the effects of grading in contrast to official Danish curricular goals, such as promoting the pupils' independence and their creativity, cooperation, and interest in lifelong learning. At the time, the first author believed that this would have an emancipatory effect leading to public knowledge about, and possible changes in, the new grade-based restricted admission to the universities. The study had no such consequence: By the time the book was ready for publication, public interest in the issue had waned. Furthermore, the book was written in an academic style, heavily documented with quotations, and contained extensive methodical discussions. I had refrained from interesting but more speculative interpretations in anticipation of the common critiques of qualitative interview research. The result was that the lived reality of the pupils' school situation was lost, and the book had no appeal to either the pupils or the general public. There were a few reviews of the book: Those in conservative newspapers were critical of the results, maintaining that they were based on too few subjects, they may have been provoked by leading questions, and the speculative interpretations were biased by the author's leftist views.

INVESTIGATING WITH THE FINAL REPORT IN MIND

The interview report is the end product of a long process; what is worth communicating to others from the wealth of interview conversations is to be conveyed in the limited number of pages of an article or a book, presenting the main aims, methods, results, and implications of an interview inquiry. The writing of the report is in this chapter presented as the last of the seven method stages of an interview study. As one approach to making interview reports more readable, we suggest taking the final report into consideration from the very start of an inquiry. In the story of the five hardship phases of an interview project, reporting was depicted as the final phase of exhaustion (Box 6.1). As a countermeasure it was recommended that an interview project be directed from the start toward the final report; that the researcher keep in mind throughout the stages of the investigation the original vision of the story he or she wants to tell the readers.

Working toward the final report from the start of an interview inquiry may contribute to a readable report of methodologically well-substantiated and interesting findings. The method steps from the original thematizing to the final report are then described in sufficient detail for the reader to ascertain the relevance of the interview design for the theme and purpose of the investigation, to evaluate the trustworthiness of the results, and, in principle, to be able to repeat the investigation.

BOX 16.1   Investigating With the Final Report in Mind

1. Thematizing. The earlier and clearer researchers keep the end product of their study in sight—the story they want to tell—the easier the writing of the report will be.

2. Designing. Researchers should keep a systematic record of the design procedure as a basis for the method section of the final report, and have the form of publishing the interviews in mind when designing the study—including the ethical guidelines of informed consent with respect to a later publishing of the subjects' stories and their potential consequences.

3. Interviewing. The ideal interview is in a form communicable to readers by the moment the sound recorder is turned off.

4. Transcribing. The readability of interviews to be published should be kept in mind when transcribing, as well as protecting the confidentiality of the subjects.

5. Analysis. The presentation of the results should be kept in mind during the interview analysis, and in some forms of analysis, analyzing and reporting merge (e.g., in narrative analysis).

6. Verification. With a conception of validation as communication and action, the reporting of a study becomes a key issue.

7. Reporting. Working toward the final report from the start of an interview study should contribute to a readable report of methodologically well-substantiated, interesting findings.

In Box 16.1, a consistent directedness toward the final report is envisaged throughout the seven stages of an interview inquiry. We regard writing as a mode of inquiry throughout an interview study. Like Richardson, who reports having been taught "not to write until I knew what I wanted to say" (Richardson & Adams St. Pierre, 2005, p. 960), we believe that writing is a central way for qualitative researchers not just to report some findings, in the final instance, but also to experiment with analyses, different perspectives on the textual material, and ways of presenting, as a method of inquiry in its own right. Writing should thus be treated as an intrinsic part of the methodology of research—and not as a final "postscript" added on.

Until recently, there has, however, been little interest in how to communicate the results of interview studies. The writing of an interview report has often been regarded as merely re-presenting what was done and found, with little regard for the readers and their use of the report. In contrast, researchers in system evaluation and market research have been well aware of the effects of the form of their reports on their intended audiences—such as the length of a report or the differential impact of quantitative and qualitative data. Patton (1980) thus mentions that an extensive, well-documented, and formally elegant evaluation report may end up in the recipient's wastebasket. A face-to-face communication, perhaps including a few pages of report summaries, may have a far stronger impact on the recipients and their decision making.

STANDARD REPORTS AND WAYS OF ENHANCING THEM

In order to evaluate the quality, validity, and transferability of the interview findings, information concerning the methodological steps of an investigation is required. The interested reader may, however, in some cases not find any, or only come across some vague scattered descriptions of how the interview knowledge was produced. One may sometimes have the impression that an interview researcher has been so insecure about his or her methodological procedures that they are preferably left unmentioned. Rather than retaining a silence on method, the unique nature of a qualitative interview study poses a challenge to give careful qualitative descriptions, as precisely as possible, of the steps, procedures, and decisions of the specific study.

Readers of an interview report can adopt a multitude of perspectives to the text: Are the results interesting, do they give new knowledge, novel insights, provoke new perspectives on the topic of the study? What are the theoretical implications of the findings? Does the new knowledge support or go against current theories in the area? From a methodic stance, questions also arise: How trustworthy are the findings? What is the methodical base for the results reported? And from a practical viewpoint, still other questions arise: What are the practical consequences of the study? Are the findings sufficiently trustworthy to act on?

In this section, standard formats for reporting interviews are outlined, and in a later section modes of enriching the interview reports are suggested.

BOX 16.2   Structuring an Interview Report

I. Introduction—*Thematizing*
The general purpose of the study is stated, the conceptual and theoretical understanding of the investigated phenomena is outlined, a review of the relevant literature on the research topic is provided, and the specific research questions for the investigation are formulated.

II. Method—*Designing, Interviewing, Transcribing, and Analyzing*
The methods applied throughout the study are described in sufficient detail for the reader to ascertain the relevance of the design for the topic and purpose of the investigation, to evaluate the trustworthiness of the results, and, in principle, to be able to replicate the investigation.

III. Results—*Analysis and Verification*
The results are reported in a form which gives a clear and well structured overview of the main findings, and with the reliability, validity, and generalizability of the findings critically evaluated.

IV. Discussion
The overall implications of the results are discussed. This involves the relevance of the findings to the original research questions and the theoretical and practical implications of the findings.

In Box 16.2, the seven stages of an interview investigation are placed under the standard headings of a scientific report: introduction, method, results, and discussion. The standard modes of reporting of the methods and the results of interview studies will now be treated in more detail.

Method

The reader of an interview report needs to know the methodical procedures in order to evaluate the trustworthiness of the results. Knowledge of specific details of method may also be required for a reinterpretation or for an application of the findings of a study. And, in rare cases, the reader may be interested in the method for replicating or extending the original study. Box 16.3 lists some of the information that a reader not satisfied with a black box in the method section can look for.

BOX 16.3   Reader Questions About Methods

Design. Selection of subjects by random, by theoretical sampling, by self-selection, or by accessibility? Such information is a precondition for decisions about applying statistical analysis to the results, and for the reader to draw generalizations from a study.

Interview Situation. What information was given to the subjects before the interview? How was the social and emotional atmosphere, the degree of rapport during the interview? What questions were posed? How was the interview guide organized? Such information is essential for interpreting the meaning of what is said in the interviews.

Transcription. What instructions were given to the transcribers, in particular with respect to verbatim versus edited transcripts? Such information is especially relevant for linguistic analyses and for psychological interpretations.

Analysis. What were the steps of the analysis? Was the analysis a personal intuitive interpretation, or were formal procedures applied? If categorizations were undertaken, how were they conducted and by whom, and how were the categories defined?

Verification. What checks were conducted of the reliability of interviewing, transcribing, and analyzing? What controls were made for counteracting biased and selective interpretations? What are the arguments for the validity of the findings?

Results

In contrast to a critic's interpretation of a literary text—where the poem or novel will be either known by or available to the reader—the interview interpreter will have to select and condense the interpreted texts for the reader.

In contrast to engaging and well-structured, rich, and "eminent" literary texts, some interviews may be boring to read, trivial, redundant, with little inner connections or deeper significance. It is up to the researcher to provide the perspectives and contexts that render the interviews engaging to the reader.

There are no standard modes of presenting the results of interview studies. There are standard ways to present quantitative data. For example, in Figure 12.1 on grading behaviors, a simple computer program provided eight graphic options for presenting the same numbers. Even though there are no comparable standard forms for presenting qualitative interview studies, there are several options available. The usual mode of presenting interview findings in the form of quotations will be treated here, and modes of enriching reports will be suggested later.

Interview Quotations. The common mode of presenting the findings of interview inquiries is through selected quotes. The quotes give the reader an impression of the interview content, and preferably also the personal interaction of the interview conversation, and they exemplify the material used for the researcher's analysis. The modes of presenting interview passages vary with the purpose of the investigation, ranging from the precise verbatim quotes used in conversation analysis (Box 10.1) to narrative restructuring (Boxes 8.3 and 13.1). In Box 16.4 we will suggest some guidelines for editing when presenting interview findings with the help of quotes.

BOX 16.4   Guidelines for Reporting Interview Quotes

1. *The quotes should be related to the general text*
    The researcher should provide a frame of reference for understanding the specific quotes and the interpretations given. The frames may vary from the lived world of the subjects to the researcher's theoretical models.

2. *The quotes should be contextualized*
    The quotes are fragments of an extensive interview context, which the researcher knows well, but which is unknown to the reader. It will be helpful to render the interview context of the quote, including the question that prompted the answer. The reader will then know whether a specific topic was introduced by the interviewer or by the subject, and possibly in a way leading to a specific answer.

(Continued)

(Continued)

3. *The quotes should be interpreted*

The researcher should state clearly what viewpoint a quote illuminates, proves, or disproves. It should not be up to the reader to guess why this specific statement was presented and what the researcher might have found so interesting in it. At the same time, the researcher's comments should not merely reproduce the quotes, perhaps in a slightly different wording, but should contribute with some perspective on the material quoted.

4. *There should be a balance between quotes and text*

The quotes should normally not make up more than half of the text in a chapter. When the interview quotes come from several subjects, each with a particular style of expression, many quotes with few connecting comments and interpretations may appear chaotic and produce a linguistic flicker.

5. *The quotes should be short*

The maximum length of an interview quote is ordinarily half a page. After half a page, the reader may lose interest, especially in an interview passage containing several different dimensions that make it difficult for the reader to find a connecting thread. If longer passages are to be presented, they may be broken up and connected with the researcher's comments and interpretations. The exception hereto is lively narrative interview passages, which may be read as stories of their own.

6. *Use only the best quote*

If two or more interview passages illustrate the same point, then use only the best, that which is the most extensive, illuminating, and well-formulated statement. For documentation it is sufficient to mention how many other subjects express the same viewpoints. If there are many different answers to a question, it will be useful to present several quotes, indicating the viewpoints they express.

7. *Interview quotes should generally be rendered into a written style*

Verbal transcriptions of oral speech, with repetitions, digressions, pauses, "hm"s, and the like are difficult to grasp when presented in a written form. Interview excerpts in a vernacular form, in particular in local dialects, provide rough reading. To facilitate comprehension, the spontaneous oral speech should in most cases in the final report be rendered into a readable written textual form. The exception is when the linguistic form is important to the study, for example in sociolinguistic and conversation analysis.

---

8. There should be a simple signature system for the editing of the quotes
The interview passages presented in the final report are more or less edited. Names and places, which break with confidentiality, will have to be altered. In order for the reader to know about the extent of editing of the quotes, the principles for editing should be given, and preferably with a simple list of signs for pauses, omissions, and the like.

These guidelines were taken from Borum and Enderud (1980) and translated, edited, and extended.

In Box 16.4, we have outlined some principles concerning how and what to quote in interview reports. It is important to emphasize, however, that these are *rules of thumb* rather than *rules as such,* and the experienced and innovative interviewer craftsman knows when these rules of thumb may preferably be broken. For example, *The Weight of the World* (Bourdieu et al., 1999), which we have referred to a number of times in this book, consists mainly of interview transcripts without many researcher comments, and this gives the book a unique style of its own that suits its purpose of throwing light on social suffering in France, seen from the perspectives of those suffering. The book has extensive reports of interviews with key phrases from the interviews interspersed as subheadings:

> the kind of redaction proposed here: breaking with the spontaneous illusion of a discourse that "speaks for itself," it deliberately works of the pragmatics of writing (particularly by adding headings and subheadings taken from phrases in the interview itself) to orient the reader's attention toward sociologically pertinent features which might escape unwary or distracted perception" (Bourdieu et al., 1999, pp. 621–622).

Other interview studies, such as those of Bellah and colleagues (1985) and Lather and Smithies (1997), also break the above "rules" in different ways. The former one by not explicitly reporting very many quotes at all (although a few are discussed in detail in the book's appendix), which gives the reader the curious impression of reading a book in empirical philosophy (the authors referred to their approach as "social science as public philosophy"), and the latter one by presenting a postmodern collage of voices, represented on the book's pages as a mixture of participants' and researchers' comments, as we will describe further below (see Box 16.5).

The Number of Pages. Quantity appears to be a persistent problem for qualitative researchers: Some seem to feel that the sheer number of pages will justify their studies not having quantitative data. Interview researchers sometimes complain that it is impossible to report the rich findings of their studies in short articles or even in books of normal size. In particular, they may want to include many pages of transcripts as documentation for their conclusions. They may also point out that it is easier to report in short form the neat designs of experimental and questionnaire studies with their quantitative data presented in simple tables and figures. The response of the editor of a Norwegian medical journal to such demands for extra pages by qualitative researchers was simple: "Everyone is special." Thus experimental researchers want more space to present the elaborate design and the sophisticated new equipment of a study, statisticians need extra space to develop the mathematical presuppositions of the statistical computations presented, and so forth.

Qualitative investigations in themselves need not require extra space—several of the qualitative studies used in the present book are in the form of short articles (e.g., Giorgi, 1975; Runyan, 1981; Scheflen, 1978). Psychotherapists may be able to present provoking findings in brief case studies, and also by means of short examples (e.g., Laing, 1962). One reason may be therapists' long experience in listening to patients, which has enabled them to become experts in attending to and selecting the essential aspects of the many stories they hear. In contrast, interview researchers who are novices in relation to their subject matter, and to interviewing, may have difficulty in developing a critical and selective distance from what they hear.

Art contains highly condensed and eloquent depictions of the manifold human condition. The quality and impact of a work of art are not necessarily enhanced by increasing its size. The short stories of Hemingway would hardly be more telling if they had been twice as long, nor would Leonardo da Vinci's *Mona Lisa* be more intriguing if painted on a double-sized canvas.

ENRICHING INTERVIEW REPORTS

No form of representation, writing, or reporting is innocent. All forms are loaded with the researchers' interests and intentions. Current developments in the social sciences have promoted an interest in the writing of research reports. A postmodern movement from knowledge as corresponding to an objective reality to knowledge as a social construction of reality involves a change in emphasis from an observation of, to a conversation and interaction with, a social world. When validation is conceptualized as a social construction of knowledge, with a communal negotiation of its meaning, communication of the findings becomes a focal part of a research project. There is a focus on cultivating the writing skills of qualitative researchers, and, inspired by postmodernism, on creative forms of writing, treating writing as a method of inquiry.

Van Maanen's *Tales of the Field* (1988) addressed the narrative conventions in ethnography for presenting the social reality of the cultures studied. From his own studies of police departments, van Maanen depicts and illustrates three kinds of tales from the field—realistic, confessional, and impressionistic.

A *realistic* tale is narrated in a dispassionate, third-person voice, with the author absent from the text. The author is "the distant one" in a realistic tale based on an assumed "Doctrine of Immaculate Perception." The natives' point of view is produced through the quotes that characterize realistic tales; the quotes render a story authentic, and the many technical and conceptual issues of constructing a transcription from an oral conversation are bypassed. With the ethnographer having the final word on how the culture is interpreted, he or she takes on an interpretative omnipotence.

The *confessional* tale, narrated in the first person, is highly personalized and self-absorbed. Mini-melodramas of hardship in the field endured and overcome, with accounts of what the fieldwork did to the ethnographer, are prominent features of confessional tales. We may today add the autoetnographic approach, where the researcher reflects on, and thereby reflexively foregrounds, his or her own experience in a given cultural situation (Ellis & Berger, 2003).

The realistic tale focuses on the known, and the confessional tale on the knower, whereas a third tale—the *impressionistic*—attempts to bring together the knower and the known by focusing on the activity of knowing. Impressionistic tales present the doing of the fieldwork rather the doer or the done. The impressionistic tale is self-conscious and, like impressionistic painting, it focuses on an innovative use of techniques and styles, highlighting the episodic, complex, and ambivalent realities studied. The impressionistic tale unfolds event by event, suggesting a learning process.

Van Maanen's (1988) goal in outlining the different styles of writing is not to establish one true way of writing ethnography, but to make ethnographers aware of the classic uses of rhetoric, such as voice, style, and audience, and from this knowledge to select consciously and carefully the voice most appropriate for the tales they want to tell.

We will now outline some ways of enhancing the readability of interview reports, suggesting genres from which qualitative researchers may learn when reporting their studies. See also Parker (2005) about more flexible structures when reporting qualitative studies.

Journalistic Interviews

One form of reporting interviews is simply to present them as *interviews.* The social science researcher may here learn from journalists, who from the start of an interview will have a specific audience in mind and usually also a limited amount of space and a nonnegotiable deadline. The newspaper or radio reporter can try to build the situation and the interpretations into the interview itself. The local context and social situation may be introduced through the interviewer's questions, for instance: "We are now sitting in the living room of the house you built when you retired, with a view through the birch forest to the fjord. Could you tell me about . . . ?" The main points and interpretations may develop from the subject's replies to the journalist's questions, or be suggested by the journalist for confirmation or disconfirmation by the subject. Thus the contextualization and interpretation can be built into the conversation, with both journalist and interviewee more or less having the intended audience in mind. The guiding line throughout the interview, the transcribing, and the editing will be to assist interviewed subjects to tell their stories as eloquently as possible to an anticipated audience.

There are exceptions to the above idealized journalistic portrait interview, such as critical, unmasking interviews. Social science is not the same as journalism, one difference being the responsibility of the social science researcher to make explicit the procedures used for editing and analyzing the interview. The journalist also has a right to protect his or her sources, which goes counter to the principle of scientific control of evidence. Yet when it comes to the presentation of results, research interviewers can still learn from good journalism.

Dialogues

Interviews can also be reported in the form of dialogues. Again, the information is conveyed through the interview interaction, but formalized and stylistically edited. Socrates' conversations with his philosophical opponents are classical examples of a philosophical discourse: All of the information is included in the dialogue, with few subsequent interpretations by the reporter—Plato. Historians of philosophy have different views on the extent to which Socrates' dialogues were direct accounts of philosophic disputes that actually took place, or whether they were mainly or entirely constructions by Plato. Independent of their status as verbatim reproductions or literary constructions, the philosophical dialogues continue to interest us today with the critical questions they pose as to the nature of truth, goodness, and beauty. Socrates' dialogues have an eminently artistic form; librarians may today have problems with whether to categorize them according to their content as philosophy or according to their artistic form as literature.

Interview findings may also be represented in the form of the researcher's dialogue with the interview texts. Hargreaves's (1994) report of his interviews with school teachers consists of theoretical and conceptual discussions of literature confronted with interview statements from the teachers' confirming, disconfirming, or refining the understanding of teachers' work in the existing literature. The continual theoretical questioning and elaboration of the teachers' statements allow for insightful and interesting reading.

Therapeutic Case Histories

A free and reflective approach to conversations and narratives can be found in reports of therapeutic interviews. Freud's clinical case stories are one illustration of an engaging and artistic presentation of conversations: He received the Goethe Prize for his writing. The works of Laing (1962) also show that it is possible through the careful use of brief conversations—theoretically interpreted from double-bind theory, psychoanalytic theory, and existential philosophy—to communicate radically new ways of understanding therapy in a simple understandable form.

In scientific psychology journals a more impersonal, formal style has generally been required. For a personal narrative of experimental studies one has to go to an exception such as the behaviorist Skinner's "A Case History in Scientific Method" (1961).

The case study is an exemplar. The use of exemplars is presently understood not as a mere popularization of theoretical points or putting some "flesh on the statistical bones" of a study. Rather, the case has its own value as an exemplar; it can, as suggested by Løvlie (1993), serve as a vehicle for learning (see Chapter 4). Donmoyer (1990) has pointed to the use of stories in teaching as a halfway house between tacit personal knowing and formal propositional knowledge. Case stories also serve as a basis for generalization in the legal and clinical fields, as discussed in the previous chapter.

The relational and tacit aspects of the interview situation are difficult to present in an explicit verbal form. The oral knowledge gained from therapy is not easily transformed into written texts. Important facets of therapeutic knowledge are best communicated by anecdotes, case stories, narratives, and metaphors (Polkinghorne, 1992). Therapists' formulation of their experiential knowledge as case stories and narratives becomes a link between the singular and the general. Such forms of transmission come closer to craftsmanship and art than to the standard norms of formal scientific reporting.

Narratives

In a narrative approach, interviewers may systematically conceive of their inquiry as storytelling from beginning to end. The report may then be a narrative rendering of subjects' spontaneous stories (see the craftman's narrative in Box 8.3), or their stories as structured into specific narrative modes, or as recast into new stories by the researcher. Scheflen's (1978) article on the interpretations of Susan's smile (Chapter 12) was cast in the narrative form of a therapeutic team watching a video recording of a family therapy session, with the therapists in turn contributing new interpretations, and with the narrator, Scheflen, weaving the threads of interpretation into a fabric. Also in the therapeutic tradition, Spence (1982) has applied narrative in both the process and the presentation of therapy.

Narratives can serve as a mode of structuring an interview during analysis (Chapter 13). The interview report itself may also have the form of a narrative rendering of the subjects' spontaneous stories, or their stories as structured into specific narrative modes, or as recast into new stories. In the latter case, the stories are reconstructed with regard to the main points the researcher wants to communicate. The narrative interviewer will encourage subjects to tell stories, assist them in developing and clarifying their stories, and during the analysis work out the narrative structures of the interview stories and possibly compose the stories to be told in the final report.

"Narrative is both a mode of reasoning and a mode of representation. People can apprehend the world narratively and people can tell about the world narratively" (Richardson, 1990, p. 21). Narratives provide a powerful access to the temporal dimension of human existence, and Richardson discusses the use of different forms of narrative reporting from everyday life, autobiography, biography, and in cultural and collective stories.

Metaphors

Novelists surpass qualitative researchers in communicating a complex social reality: "Their appeal is that they dramatize, amplify, and depict, rather than simply describe social phenomena. The language itself is often figurative and connotative, rather than solely literal and denotative. Part of this has to do with the use of metaphors, analogies, symbols, and other allusive techniques of expression" (Miles & Huberman, 1984, p. 221). A study's main points may be more easily understood and remembered when worked into vivid metaphors. Through a metaphor, one kind of thing is understood in terms of another. Psychoanalysis is replete with metaphors, often taken from myths and literature, such as the Oedipus and the Electra complexes. Metaphors also, though often unnoticed, permeate mainstream social scientific writing with terms like theory-"building," knowledge as "enlightening," and so on.

Miles and Huberman (1994) advocate the use of metaphors in reporting qualitative studies. A metaphor is richer, more complete than a simple description of the data. Metaphors are data-reducing and pattern-making devices. Miles and Huberman thus found in a school improvement study that a remedial reading room felt like an "oasis" for the students sent there. The metaphor "oasis" pulls together separate bits of information: The larger school is harsh and resource-thin, like a desert, and some resources are abundant in the pupils' remedial room, like the water in an oasis.

Visualizing

Although interview data are of a verbal nature, the possibilities of presenting the results in visual form should not be overlooked. Quantitative data are today often presented visually in the form of graphs and figures. A comparative choice of standard visual modes of presentation does not exist for qualitative inquiries. There are, however, several options, such as a tree graph of the main categories and their subcategories, diagrams with boxes and arrows showing the main sequences of a story, and the like.

If the researcher has artistic abilities, interview results may be presented as drawings. In a different field, a Danish professor of architecture found that he could not convey to his students through lectures or writing what he found essential about modern and postmodern architecture. He then resorted to collages, where through placing buildings in unexpected contexts and from new angles, such as an opera house under the sea, he was better able to comment on the current situation of architecture than through words (Lund, 1990). The collages attained an aesthetic value of their own, and have been displayed in art galleries.

Collage

We conclude with an interview report breaking new ground: *Troubling the Angels: Women Living With HIV/AIDS* (Lather & Smithies, 1997). The researchers talked with HIV positive women, in their support groups and individually, charting the journey of their struggles from infection to symptom, to sickness, to wasting, to death. The book is organized as layers of various kinds of information, where the major part consists of conversations in support groups for the women, with the researchers entering with questions and comments at times. The style of reporting remains faithful to the messy, polyphonic character of postmodern life. The stories of the women are interspersed with inter-texts on angels, chronicling the social and cultural issues raised by the disease. Factual information on HIV/AIDS is presented in boxes throughout the text. Across the bottom of much of the book is a running commentary by the co-researchers, moving between research methods and theoretical frameworks to the co-researchers' autobiographies throughout the inquiry, with their reactions to the research theme.

BOX 16.5   Reporting Interviews With Women With HIV/AIDS

Ana: I want to comment on that real quick. I took a friend of mine to get tested and this person turned out to be negative and the guy said you're lucky you're negative, because if you were positive you would die. My friend didn't know that I was positive.

Geneva: I was 20 when I found out. I was going through the nursing program at the community college. I had to go through Job Corp for them to pay for it and I had to take all these tests, including a mandatory AIDS test. So when they called me up, first I waited in this clinic for a good hour, me and three guys. Then I started panicking because I knew they had done an

AIDS test. Then they had us each come in and told us we were positive. It was really cold, kind of "legally we have to inform you." I was crying so hard. The way you find out can be devastating.

Maria: I wanted to mention real quickly how I found out. This was back in '87. It was my birthday and I thought it would be nice to find out that I was negative. I didn't want to drive back to the testing place, so I called. And there was one particular person I asked for and I said, "It's my birthday and I don't want to drive into the city can you just tell me?" He goes, "I can't do that, you have to come in." I was pleading so he told me over the phone. And I just stayed quiet for a minute. And I was at work and I was looking around and all of a sudden I thought I was going to pass out.

* * * * * * * *

Patti: In a September 30, 1993, *Rolling Stone* interview, Randi Shilts, author of *And the Band Played On,* revealed that he found out he was HIV positive as he finished writing the book. I knew early on that my own HIV status was an important factor in my positioning in this study. I was tested on August 12 of 1993 and received the results of my negative status on September 8. Going through the tension of getting the test and, especially, waiting to hear the results, I learned that part of me didn't want to know if I was positive, but that I wanted very much to know if I was negative. It meant that I was "safe" somehow and could stop worrying that every skin spot, every tiredness was some sign, a situation I hadn't given much thought to until I had the test. The most important thing was to realize how easily I could have been positive, how much of it is sheer luck. It's like a poker hand you get dealt, caught up in history's net, one way or the other. I remember thinking that if I were negative, I would be grateful, and I promised I would be CAREFUL, if allowed to escape this time, this 1993, at the end of this first decade of the epidemic.

If positive, I thought of how my identity would shift and the world would be so very different, including the perspective I would bring to this project. I would be much more like Francisco Ibañez-Carrasco in his 1992 study of gay men in Vancouver and Chile: one of, studying across, instead of "down" or, as I seem to be doing "up." Or maybe not. . . .

Reflexive Coda: Confessional writing is not my cup of tea, but I was fairly satisfied that the preceding was useful in moving toward some emotional shape for this work.

SOURCE: *Troubling the Angels: Women Living With HIV/AIDS* (pp. P. Lather & C. Smithies, 1997, Boulder, CO: Westview Press. R permission of Westview Press.

Box 16.5 shows the layout of one page from Lather and Smithies' book, in this case reporting at the top the stories of the women interviewed about learning how they had gotten AIDS, and at the bottom the reactions of one of the researchers to receiving the results from her own test.

The authors hoped with the book to provide support and information to the women and their friends and families and to promote public awareness of their issues. They had promised the women to be the editorial board for their book and prepared a desk copy version for the women's member checks. They participated eagerly in the process and could be frank with their impatience "Where is the book? Some of us are on the deadline, you know!" Challenging any easy reading via shifting styles and multiplication of layers of meaning, the final book attempts to do justice to the women's lives, positioning the reader as thinking, willing to trouble the taken-for-granted, opening to that which is beyond the word and the rational.

PUBLISHING QUALITATIVE RESEARCH

Fortunately, most disciplines in the social sciences today have their own qualitative journals, where it is possible to publish interview reports. There are outlets that specialize in qualitative studies such as *Qualitative Inquiry, Qualitative Research, Qualitative Sociology, Qualitative Research in Psychology, International Journal of Qualitative Studies in Education,* and *Qualitative Health Research,* in addition to the more general journals that publish manuscripts based on quantitative as well as qualitative research.

Still, it may be hard for qualitative researchers to have their reports accepted in the "top journals" such as *Psychological Review* or *American Sociological Review.* Richardson (1990) describes how she finally got a qualitative research paper based on interviews accepted in the latter journal, and her advice may still be useful today. First, one can preferably identify oneself with a particular theoretical tradition and methodological approach, and, in her case, Richardson chose symbolic interactionism, which is a classical tradition in American sociology, not too alien to the postmodern interests of the author. Second, she made only limited claims for the statistical representativeness of her findings, but unlimited claims to their generality and theoretical significance. This she could do by, third, working with typologies (in the tradition of Max Weber's ideal types) that grew out of her empirical material, which can be valid as qualitatively based categories even without quantitative information concerning how the general population is distributed in the categories. Fourth, her methods section was quite long, describing in detail what she had done and how she had chosen her quotes (e.g., she made clear that she had only selected quotes that represented common themes). Qualitative researchers may profit from reading Richardson's descriptions, and may possibly become inspired to try for themselves to aim at one of the "top journals" in their field, although, today, there are a number of outlets for more "unconventional" ways of reporting qualitative research, such as *Qualitative Inquiry* and the *International Journal of Qualitative Studies in Education.*

SEVENTEEN

CONVERSATIONS
ABOUT INTERVIEWS

---◆•●•◆---

T hroughout the book we have treated the interview as a form of conversation. We now turn to conversations with other researchers and users of
interview research about the value and quality of the knowledge produced by
qualitative research. We start with some internal critiques of the quality, credibility, and value of the knowledge that is currently produced by research interviews. Then we sum up suggestions from earlier chapters, and discuss ways of
enhancing the quality of interview research along three lines. We first discuss
how to learn and apply the craft of interviewing within an academic setting.
Then we return to the epistemological outline of interview knowledge presented earlier, and discuss how to conceptualize the knowledge produced by
qualitative interviews in order to preserve and strengthen the richness of this
knowledge. Finally, we address qualitative research interviewing as a social
and ethical practice, situated in a specific cultural context.

CRITIQUES OF
THE QUALITY OF INTERVIEW KNOWLEDGE

Learning interview research goes beyond a technical mastery of the interview
craft to include professional reflection on interview practice and on the value of
the interview-produced knowledge, with an awareness of the epistemological

and ethical issues involved. Interview research may be controversial, and different audiences may judge the value of an interview inquiry rather differently, depending to some extent on their implicit epistemologies. Some often-repeated mainstream objections to interview research were presented in Box 9.3. Today, dissatisfaction in a different vein comes from audiences sympathetic to the idea of qualitative interview research, such as scholars in the humanities and also professionals. These critics are closer to an understanding of the interviewer as a traveler returning home with tales from conversations in a new country.

BOX 17.1   Internal Criticisms of Interview Knowledge

Current interview research is:

- *Individualistic*—it focuses on the individual and neglects a person's embeddedness in social interactions
- *Idealistic*—it ignores the situatedness of human experience and behavior in a social, historical, and material world
- *Credulous*—it takes everything an interviewee says at face value, without maintaining a critical attitude
- *Intellectualistic*—it neglects the emotional aspects of knowledge, overlooks empathy as a mode of knowing
- *Cognitivist*—it focuses on thoughts and experiences at the expense of action
- *Immobile*—the subjects sit and talk, they do not move or act in the world
- *Verbalizing*—it makes a fetish of verbal interaction and transcripts, neglects the bodily interaction in the interview situation
- *Alinguistic*—although the medium is language, linguistic approaches to language are few
- *Atheoretical*—it entails a cult of interview statements, and disregards theoretical analyses of the field studied
- *Arhetorical*—its published reports are boring collections of interview quotes, rather than convincing stories
- *Insignificant*—it produces trivialities and hardly any new knowledge worth mentioning
- *Legitimizing*—it is obsessed with legitimizing itself rather than producing new knowledge

Some of the insider objections to current interview research are presented in Box 17.1. The objections primarily concern the nature of the knowledge that qualitative interviews produce. Suggestions for overcoming some of these critiques have been put forth in previous chapters. Below, we briefly review some ways of enhancing interview research in order to demonstrate that the criticized features do not necessarily pertain to interview inquiries as such, but rather to what have been common modes of applying and understanding interviews in research.

Most interview research today is carried out as individual interviews and may lead to a methodological individualism. The use of group interviews, such as focus groups, however, can bring up lively *interpersonal dynamics* and show the social interactions leading to the interview statements (see Barbour, 2007). Interviews can be used to obtain descriptions of the cultural and the historical, the social and the material, contexts of subjects' lives. Anthropological interviews focus on the respondent's culture, and interviews can be employed in recording oral history. Some interview studies that have focused on the *material and social situations* of the subjects include Oscar Lewis's descriptions of Mexican peasants' life situation in *The Children of Sanchez* (1964), as an older example, and Bourdieu's study of the plight of the downtrodden in France in *The Weight of the World* (1999) as a recent example. Also, the attempt of the first author of this book to link the effects of learning for grades to the requirements of wage labor, which we have referred to throughout the book, can be mentioned as an example of relating interview statements to their broader social and economical contexts.

Some interview researchers practice an active approach to interviewing, allowing themselves to question and challenge the respondents' answers, thereby countering the *credulous attitude* of many interview studies. The use of therapeutic interviews for research shows the possibilities of applying *empathy and emotional interaction* to obtain significant knowledge of the human situation. In recent feminist qualitative research there is an emphasis on the knowledge potentials of feelings, empathy, and the personal dimension in human interaction, including research interviews (Ellis & Berger, 2003).

There are trends today toward giving the knowledge obtained through interviews back to the participants in the social situation in which the knowledge was developed. This pertains to the use of interviews in system evaluation; and in action research and feminist research particularly, the knowledge acquired through interviews is often utilized pragmatically to *change* the situations investigated. Though most interview research today is chair-bound, researchers might learn from radio and TV interviewers, who sometimes walk around with their subjects in the subjects' natural surroundings, such as their workplace or home. *Conversations integrated into the subjects' natural activities* of their daily world provide a more comprehensive picture of their background situation than the office-based views.

The verbal fixation of interview research may, to a certain extent, be counteracted by video recordings of interviews, whereby one retains access to the *bodily expressions* and interpersonal dynamics of the interaction. The development of computer programs for directly analyzing the audio and video recordings allows a move from the alienation of transcripts to the listening to conversations. Parallel to the trend of including nonlinguistic aspects of interview conversations is a movement toward going beyond the naive use of language in social science interviewing through an orientation to *sociolinguistics* and the pragmatics of speech. This includes an increasing awareness of the differences between the oral speech of the interviews and the written texts analyzed. Discourse analysis, ethnomethodology, and conversation analysis have contributed to heightening researchers' attention to the pivotal role of language in interview research.

The impact of Glaser and Strauss's grounded theory in current qualitative research shows the possibility of formulating *theories* grounded in empirical interviewing. Within the psychoanalytical tradition, the development of innovative theories on the basis of patient interviews has not been limited to Freud's original contributions at the turn of the century, but has continued with theories originating with therapeutic interviews about the authoritarian personality in the 1930s and 1940s, and the narcissistic personality in the 1970s and 1980s. The *communication of qualitative studies* has come to the fore in the past few years, drawing on narrative approaches and the rhetorical tradition, and also through experiments with artistic forms of reporting interview studies (e.g., the forms of poems and short stories).

Regarding the criticism concerning *insignificant interview findings,* the often boring reading of current interview reports, as discussed in Chapter 16, may not only be a matter of the rhetoric of reporting, but may also concern the very results of an interview study. There may be a lack of substantial findings in some published interview reports, a lack of significant new knowledge in relation to common sense or established scholarly knowledge. In contrast to the extensive contributions of psychoanalytic interviews and sociological and ethnographic field studies, it is harder to point out significant new bodies of knowledge that have come from current qualitative interview research. According to one verdict, qualitative researchers often produce "low quality research and research results that are quite stereotypical and close to common sense" (Seale, Gobo, Gubrium, & Silverman, 2004, p. 2).

There exists a certain contrast between visions from the 1970s of a radically new research method in the social sciences and the amount of significant new knowledge produced today by qualitative interviews. Many textbooks on qualitative paradigms and methodologies have appeared, but few breakthroughs in knowledge. The interview studies mentioned in this book by Bellah, Bourdieu, Hargreaves, Sennett, and Schön, based on an extensive theoretical preknowledge of their interview themes, are exceptions in this regard. Several of these studies have been influential not just in narrow social science circles, but also in a wider societal context; *The Weight of the World* (Bourdieu et al., 1999), for example, has sold more than 140,000 copies in France alone. Furthermore, these studies remain—in contrast to current interview studies with often extensively paradigmatic legitimations of methodology—in general tacit on the varieties of qualitative paradigms, and give rather short depictions of the methods used. In both *Habits of the Heart* (Bellah et al., 1985) and *The Weight of the World* (Bourdieu et al., 1999), the methodological sections are limited to rather short (but helpful and readable) appendices. The major emphasis in these studies is on the interview findings and their theoretical and practical implications. Perhaps we encounter in current interview research a negative correlation between number of pages on methodological paradigms and number of pages contributing with substantial new knowledge in a field. According to one review of a leading handbook in the field: "Qualitative research, as presented in the *Handbook* [by Denzin and Lincoln], appears to thrive on perennial crisis, on legitimization, of praxis and of representation. There is an abundance of theorists, but a dearth of practitioners; a preponderance of assertions, but little evidence or rational argument in support of them" (Ho, Ho, & Ng, 2007, p. 382).

We agree that this latter line of criticism applies to some interview research, but, as we have argued throughout this book, we also believe that the development of high quality interview research goes hand in hand with a reflection on its theoretical presuppositions and implications. Enhancing the quality of knowledge production in research interviews is not merely a question of improving the research skills of the individual researchers, nor of enlarging the methodological scope of ways of interviewing. Enhancing interview quality is also a conceptual issue, which—beyond the theoretical knowledge of the subject matter of the investigation just mentioned—also entails researchers' awareness of epistemological and ethical issues raised by research conversations about private lives for public consumption.

DEVELOPING THE CRAFT OF RESEARCH INTERVIEWING

Research interviewing is a craft and not a systematic method with more or less mechanical rules to follow. Interviewing is a skill to be learned through own practice, by watching masters of the trade interviewing, and by reading eminent interview reports. A craft conception of interviewing involves no relief from the rigor of method, but places the exertion of rigor in the levels of skills and the personal judgments of the interviewer. While a craft cannot be learned through reading a book, we have attempted to describe the craft of interviewing, provided examples and rules of thumb, and suggested learning tasks to help newcomers become skilled at interview research by doing interview research. When we have exemplified interviewing with reference to historical masters of the trade, it may have scared some readers by putting the quality level of interviewing at an unattainable level, not reachable within the short span of a masters or Ph.D. program. The examples of master interviewers, such as Socrates, do not serve to set the level of common research interviewing, but are presented as ideals to learn from.

Interviewing produces knowledge, and sound interviewing involves a conceptual grasp of the subject matter of interviewing as well as of the kind of knowledge produced by interviews. Throughout the book we have given examples of how concrete practical decisions on how to go about interviewing depend on conceptions of the knowledge one wants to obtain through interviewing. Our emphasis on the relevance of conceptual knowledge for practicing the craft of research interviewing does not imply that interview researchers in their reports need to take a long tour through the many paradigms that currently legitimate the qualitative field; beyond providing an elementary familiarity with epistemology, they should instead stick to their research projects and reflect on the specific forms of knowledge they aspire to produce. Such concrete reflections on the hows of knowledge production may enhance the quality of interview research.

Writing a book on the craft of interviewing thus involves the difficulty of presenting explicit and general guidelines for a craft that consists of practical skills and implicit personal know-how. A related paradox emerged when we discussed ethics in Chapter 4, where we argued that explicit ethical guidelines and rules never dictate their own application, but likewise depend for their use on practical skills and know-how. Applying rules and principles is a concrete and situated affair, demanding proficient judgment concerning the particular. While one does not become a good interviewer or an ethically good person through reading a book about interviews or ethics, a book can nevertheless provide information about the terrain through which the journey goes and potential dilemmas on the way, and about available conceptual and technical equipment for the journey, and thereby facilitate the journey and enhance the quality of the knowledge that the interview traveler brings home. When we have drawn in the examples of significant studies in the social sciences based on interviews, we have had the aim of looking over the shoulders of capable interview researchers. Learning the practical skills of interviewing, as well as learning ethical behavior, takes place in practice, through *doing,* and studying exemplars of excellent interview research is also important, just as the novice architect can learn much from visiting the great buildings of a culture. Research interviewing involves a cultivation of conversational skills that most of us possess in advance. However, even though most human beings can open their mouths and sing, for example, only few of them sing well enough to attract the attention of other people, making them interested in listening. Even fewer sing well enough that others are willing to pay to listen to them for hours.

It is hard to produce new knowledge through interviews that goes beyond common sense and which may be pragmatically helpful in understanding or even changing a social situation. Some interview research can be characterized as "tourist interviewing," where the interviewer, without substantial prior training, meets the subject for a very brief period of time, records a conversation in snapshot mode, and subsequently tries to say something interesting about it. In contrast, as we have pointed out, interviewers who make a living based on their craftsmanship, such as focus group market researchers or therapists, not only engage in intensive and extensive interview training, but also—especially therapists—must become closely acquainted with their subjects over a long period of time in order for the subjects to say meaningful things about their lives (this also applies to anthropological fieldwork). Such professional conversationalists represent the antithesis to "tourist interviewing."

A practical problem for the conception of research interviewing as a craft concerns how to obtain legitimacy for a craft within a method driven academic culture. When applying for a Ph.D. or a research grant, one often has to pretend that research is a well-ordered process that can be codified into discrete methodical steps (see Box 6.5), an impression that also partly colors this book, with its depiction of the seven stages of an interview project. Although such steps and procedures may be valid and helpful as analytic tools, they represent a process that is much more fluid and dynamic than such discrete and linear categories reveal. Many interview studies are not designed once and for all, prior to the actual interviews, for example, since often researchers become wiser along the way, leading to a partial redesign of the project. As we have emphasized, the analyzing stage is preferably initiated already in the course of face-to-face interviewing; this also pertains to the verifying stage, where the interviewer can try to test his or her interpretations in the conversation itself. Likewise, reporting is not a discrete stage at the end, for writing is a method of inquiry that is preferably undertaken throughout the project. The production of new and insightful knowledge through interviews owes more to the role of researchers as persons, their skills, capabilities, and intellectual virtues, than to their methods in a mechanical sense. The quality of the knowledge produced through interviews further rests on an understanding of the nature of interview knowledge, to which we turn now.

AN EPISTEMOLOGY OF INTERVIEW KNOWLEDGE

Throughout this book, we have argued that qualitative research can lead to valid descriptions of the qualitative human world, and that qualitative interviewing can provide us with valid knowledge about our conversational reality. Research interviewing is thus a knowledge producing activity, but the question is how to characterize the form of knowledge that qualitative research interviewing can give us. In Chapter 3 on epistemology, we presented the metaphors of knowledge as a given substance to be collected by a miner and knowledge as socially constructed, entailing a conception of the interview researcher as a traveler, composing stories about the journey. Qualitative research interviewing can rely on both metaphors, with Glaser and Strauss' (1967) grounded theory approach as an example of the former, advocating inductive theory building through the collection of data that exist independently of the method and the researcher.

Social constructionist approaches, such as in discourse analysis, represent the latter view that data are not given, but created in and through the conversation, leading to a conception of the interview as a meaning-making practice (e.g., Holstein & Gubrium, 1995).

We have argued that different epistemologies are not just different icings on the same cake, for the practical decisions that interview researchers have to make in their projects are implicitly or explicitly informed by their conceptions of knowledge. Presupposing that knowledge emerges from a collection of given data will naturally lead to a view of leading questions as a bias in qualitative interviewing, detrimental to the process of acquiring objective knowledge. And presupposing that knowledge is socially constructed can lead to a view of leading questions as one way of inquiring into the strength and justifiability of a subject's beliefs, perhaps even yielding another form of objectivity in the sense of provoking the object to object, thereby letting it show more interesting sides of itself. A miner's epistemology will lead to a "pipeline approach" to the interview, regarding the conversation as a neutral channel through which information is moved, whereas a traveler's epistemology will lead to a view of the conversation as a construction site of knowledge. The production of knowledge through interviews can then take different forms, a perhaps extreme version is the Socratic approach, where knowledge in the sense of *episteme* results only when beliefs can be justified discursively (e.g., as a result of an active and sometimes challenging confrontation between conversation partners).

In Chapter 3, we presented interview knowledge as something *produced,* constructed in the interaction of interviewer and interviewee; *relational,* arising through concrete human relations; *conversational,* arrived at through questions, answers, and descriptions; *contextual,* with the meanings more or less tied to specific contexts; *linguistic,* carried in the medium of spoken and later written language; *narrative,* disclosing the storied nature of the lived human world; and *pragmatic,* ultimately deriving its legitimacy from enabling us to cope with the social world in which we find ourselves. We will now summarize some of the epistemological points that have been discussed in the book in relation to these intertwined features of interview knowledge.

Produced Knowledge. In contrast to a view of interviewing as collecting knowledge for later analysis by appropriate computer programs, the approach of this book has gone beyond a prevalent separation of data collection and data analysis. Interview knowledge is not collected, but produced between interviewer and interviewee, and the meanings constructed in their interaction are again restructured through later stages of an interview inquiry. An emphasis on knowledge as something produced throughout the interview inquiry not only brings into focus the quality of the original interview situation, but also takes account of the constructive character of transcripts, as well as efforts to enrich the knowledge production when reporting interviews to the readers.

Relational Knowledge. In this book, we have conceived of research interviews as literally inter views, that is, not as an interviewee monologue recorded by an interviewer, but as a co-constructive process involving both parties. Thus, the relation and the personal interaction between interviewer and interviewee are central to the interview, which means that different constellations of interviewers and interviewees will result in different knowledge products. The able interview craftsman will create a relationship that leads to the production of significant knowledge. And different forms of interviewing involve different personal relationships, which lead to different kinds of knowledge. The interview, therefore, is not a standardized and mechanical approach, but one where the unique relation between human beings determines the result.

Conversational Knowledge. We live in a conversational world. The relevance of conversations for social science goes beyond the use of interview conversations as an additional empirical method. It includes conversations among researchers and the public about the truth and value of the knowledge produced in interview conversations. From an understanding of objectivity as letting the object speak and object, the qualitative research interview obtains a privileged position for producing objective knowledge of a conversational world. According to the perspective laid out in this book, conversations are both a set techniques, a mode of knowing, and related to a fundamental ontology. First, the research interview can be treated as a specific professional form of *conversational technique* in which knowledge is constructed through the interaction of interviewer and interviewee as outlined in the description in Chapter 2 of the mode of understanding in the qualitative research interview. In contrast to the reciprocal interchanges of everyday life, as well as of philosophical conversations, it is the interviewer who, as a professional, asks and the interviewee who answers. Second, the conversation may be conceived of as a *basic mode of knowing.* The certainty of our social knowledge is a matter of conversation between persons, rather than a matter of interaction with a nonhuman reality. Third, *human reality* may on an ontological level be understood as persons in conversation. We are conversational beings for whom language is a fundamental reality. The conversation is not only a specific empirical method: It also involves a basic mode of constituting knowledge and a view of the human world as a conversational reality.

Contextual Knowledge. Interview knowledge is produced in a specific interpersonal situation, and the situational and interactional factors influencing the knowledge produced need to be taken into account, which is done today, for example, in discursive and conversation analyses of interviews. What matters is not arriving at context-independent universal knowledge, but producing thick descriptions of situated knowledge from the interviews. The transfer value of this knowledge to other situations may then be critically evaluated by other researchers and by lay readers. Similarly, following Aristotle, we argued in Chapter 4 that ethical knowledge is basically contextual and situated— knowing how to act well in the here and now cannot be captured in general and universal rules or principles.

Linguistic Knowledge. In line with an understanding of interviews as a specific form of conversation, the skilled interviewer needs to master the medium of conversation—language. For a professional analysis of the interviews and their transcripts, the researcher should be familiar with linguistic tools for analysis of language. Social scientists working with numbers as a rule employ a variety of professional tools for the analysis of numbers, such as statistics. In graduate social science programs, courses in statistics tend to be mandatory and courses in linguistics nonexistent. In order to reach a professional level comparable to that of quantitative analysis today, qualitative social research needs to enhance the quality of interview research by moving beyond a linguistic illiteracy toward a professional mastery of the linguistic medium of the interview craft.

Narrative Knowledge. People's tendency to tell stories about their lives and experiences should not be regarded as a subjective distortion of objective facts. Rather, important parts of human living and experiencing are storied, which means that a narrative expression of these parts is needed in order to capture their essential features. Interviewers may take advantage of this and encourage interviewees to tell stories, and interview researchers may, when they write up their reports, stick with the narrative form and convey their material as stories, originally co-constructed in the interaction of researcher and participant. Thus, a narrative approach can be followed from the interviewing stage, through the analysis, and to the reporting of interviews.

Pragmatic Knowledge. Throughout this book there has been a pragmatic emphasis on learning interviewing from practice, not only from one's own interview practice, but also from the practices of historical interview studies, which have led to significant differences in the way we understand social phenomena. This implies a move from interview research as methodological rule following, with method as a truth guarantee, to research as craft, where craftsmanship is learned through practice, and the value of the knowledge produced is the key quality criterion.

We have taken a pragmatic stance toward both research interviewing and the knowledge it produces. Pragmatists believe that trees shall be known by their fruits, and we believe that the best legitimation of qualitative interviewing as a form of research is found in the quality and value of the knowledge produced through interviewing, rather than in subtle paradigmatic discussions. This explains our many references to the interview studies that have been significant in the social sciences in the last 20 years (e.g., those of Bellah, Bourdieu, Hargreaves, Schön, Sennett, etc.). According to pragmatism, humans can be said to possess knowledge when they are able to act in desired ways; knowing is a kind of doing, a doing which, in Dewey's eyes, is based more on prereflective habits than on conscious reflection (Dewey, 1922/1930; Brinkmann, 2004). This pragmatic view of knowledge as an activity goes hand in hand with the craft conception of interviewing, according to which the production of knowledge through interviews is primarily a practical matter of acquiring skills through hard "doing."

A pragmatic approach involves a transition from philosophical legitimation to the practical effects of knowledge. Today, we may discern a move from questions about objectivity and validity of interview knowledge to questions about the quality and value of the knowledge produced. In contrast to general paradigmatic questions—such as "Is the interview a scientific method?" or "Is the knowledge from qualitative interviews objective?"—pragmatic questions are specific and practical—such as "Is this interview-produced knowledge useful?" and "Useful for whom and for what?"; "Are these results worthwhile, valuable, insightful, and beneficial?"; "Are the findings sound?"; "Are they worth depending on?" The question of application raises the issue of useful and beneficial for what purposes and to whom, leading to ethical and sociopolitical questions about the use of the knowledge produced. The quality of interview knowledge is not just an issue of methodology or epistemology, but also of ethics. Before we turn to interviewing as a social practice, we will briefly address the primacy of content over method in interview research, and a prevailing supremacy of a quantitative language.

THE OBJECT DETERMINES THE METHOD

Above we have attempted to outline key features of the knowledge produced by qualitative research interviewing. These features also characterize the object, or subject matter, of qualitative research as the *produced* (or constructed), *relational, conversational, contextual, linguistic, narrative,* and *pragmatic* (or action-oriented) parts of reality. Throughout the book, we have maintained that the nature of the subject matter enjoys primacy, and should dictate which methods to use. We should consider *what* we want to know before determining our *ways* of knowing it. Here, we can refer to the Danish statistician Karpatschof, who has argued that the choice between quantitative and qualitative methods should be grounded in the nature of the subject matter of interest (Karpatschof, 2006). Inspired by Sartre's distinction in the *Critique of Dialectical Reason* between *groups* and *series,* Karpatschof argues that there is a fundamental ontological difference between phenomena that exist contextually, in social groups with complex social interaction (e.g., in a group working together), and phenomena that exist as part of individual behavior in series, where people act in parallel ways, but independently of each other (e.g., when waiting in line at the bus stop). Phenomena characterized by *seriality* are atomized and personal individuality is unimportant, whereas uniqueness is important for phenomena characterized by *contextuality* (Karpatschof, 2007). Karpatschof's point is that quantitative methods are suitable for investigating serialized phenomena (i.e., those aspects of persons that are salient when they are regarded as parts of series), and qualitative methods are suitable for investigating contextualized phenomena, where persons are considered as members of social groups in Sartre's sense.

Karpatschof further recounts the story of how societies have become increasingly serialized with the transition from premodern societies, organized through organic group relations, to modern bureaucratized societies, a transition that created a need and a possibility for statistical information about the population. Statistics originally developed as a "science of the state" with the purpose of governing populations (see Rose, 1996). From this perspective, it is no coincidence that ethnographers often work primarily with qualitative methods (since it is often societies with little seriality that are the subject matter of ethnographic studies), whereas sociology has been a much more quantitative discipline based on statistics with the study of modern, highly serialized societies. From this historical perspective, we may add that qualitative methods again come to the fore in the increasingly deregulated consumer societies of a postmodern age, centered on an experiential economy, which require contextualized qualitative methods of inquiry.

On the basis of the ontological distinction between series and groups, Karpatschof concludes that in psychology quantitative methods are suitable for the study of basic perceptual phenomena and serialized societal phenomena, but all other areas demand the use of qualitative methods due to the concreteness and nonseriality of the subject matters (Karpatschof, 2006, p. 48). Harré (2004) has gone even further and argues that since psychological and other social phenomena are collective discursive acts, meaningful performances by skilled human actors ("group phenomena" in the words of Sartre and Karpatschof), qualitative researchers should simply dismiss the objection from mainstream psychology that their methods are "unscientific." They should point out that "it is the qualitative techniques and the metaphysical presumptions that back them that come much closer [than mainstream psychology] to meeting the ideals of the natural sciences." (p. 4) (see also our discussion in Chapter 15 of objectivity as being adequate to the object). According to Harré, qualitative methods are scientific because they are adequate to investigate a qualitative human world. If this is true, however, then we need to ask why so many researchers and funding agencies still advance a "quantitative imperative"? We now turn to this question.

THE SOCIAL SCIENCE DOGMA OF QUANTIFICATION

A demand for quantification of research material has dominated the social and health sciences, whereby qualitative approaches have tended to be ruled out, or relegated to a lower scientific status. A widely used textbook on social science research expressed a representative view on quantification as follows:

> Scientists are not and cannot be concerned with the individual case. They seek laws, systematic relations, explanations of phenomena. And their results are always statistical. . . . the existential individual, the core of the individuality, forever escapes the scientist. He is chained to group data, statistical prediction, and probabilistic estimates. (Kerlinger, 1979, pp. 270, 272)

A leading psychologist has put it more succinctly—"one's knowledge of science begins when he can measure what he is speaking about, and express it with numbers" (Eysenck, 1973, p.7).

The responsibility for the quantitative rule in the social sciences is often given to positivism. While quantification of knowledge was a key element of methodological positivism of mid-20th century, the social science dogma of quantification was not an original part of positivist philosophy. Comte, the founder of classical positivism, believed that social phenomena were too complex to be subjected to mathematical analysis, and the application of mathematical analysis was not at all necessary for a positive science:

> our business is to study phenomena, in the characters and relations in which they present themselves to us, abstaining from introducing considerations of quantities, and mathematical laws, which is beyond our power to apply. (Comte, 1975, p. 112; quoted from Michell, 2003, p. 13)

The Social Scientist's Natural Science Comedy of Errors. The quantitative imperative of the social sciences has often been justified by pointing to the natural sciences as quantitative sciences. This is pure myth, however, for the research practices of the natural sciences have long been of mixed methods, where qualitative and quantitative analyses are both useful tools; in some instances the major evidence is of a qualitative nature (e.g., in Darwin's theory of evolution). Thus, empirical qualitative researchers need not necessarily believe what some philosophers and social scientists have written about the methods of the natural sciences, but could instead interview natural scientists on campus about their research practices. Or they may go to the university bookstore and ask for a book on qualitative analysis, which can just as well lead us to the chemistry shelves as to the social sciences. Or, even more simply: At the time we finished this book (October, 2007), a Google search on "qualitative analysis chemistry" yielded more than 2 million hits and "qualitative analysis social science" 6.3 million hits (with "quantitative" in the search, the hits were over 17 million and 8.5 million). A thorough study of texts on physics concluded that "the physics of the physicist" and "the physics of the psychologist" are two entirely different worlds (Brandt, 1973). An anthropologist, who has actually entered into and observed research behavior in natural science laboratories, concluded laconically: "The imitation of the natural sciences by the social sciences has so far been a comedy of errors" (Latour, 2000, p.14).

A Religious Quantification. We thus need to go beyond positivist philosophy and the natural sciences to find the origins of the quantitative dogma of social scientists. As with many important phenomena in Western culture, we may start with the Bible—God "ordered all things by measure, number and weight" (from the Book of Wisdom 11:21). Michell (2003) has traced the quantitative imperative to this saying of the Bible, and to the Greek mystical religion of numbers by the Pythagoreans, who maintained that everything that exists, exists in numbers. St. Augustine connected the Pythagorean quantitative cosmology of Plato in Timaeus with the Bible, concluding that "supreme measure, the supreme number, and the supreme order" are attributes of God. St. Augustine's dictum that the quantitative structure of the world reflects the Divine nature of the world came to dominate medieval thought (Michell, 2003, p. 11). When Galileo a thousand years later advocated: "measure what is measurable and make measurable what is not," this was not against the Catholic Church, but a radicalization of St. Augustine's divine quantitative cosmology.

Quantitative Funding. Locating the origin of the quantitative dogma of science in Greek and Christian religion does not explain the current prevalence of the quantitative dogma of contemporary social scientists. Michell (2003) traces the dominance of the quantitative imperative from the middle of the 20th century, when it was advocated by psychometric psychologists such as Stevens, to the economic structures supporting research in psychology. In a unified methodological position based upon quantification, the references to positivist philosophy and the natural sciences served as window-dressing, helping the funding of psychological research.

The social science dogma of quantification is today in line with a rationalization and bureaucratization of society, conceptualized as the current audit culture. It conforms to the all-pervasive economic performativity of a postmodern capitalist economy, where everything that exists can be measured in money, or dismissed—in the words of Lyotard (1984, p. xxiv): "be operational (that is, commensurable) or disappear." We may conclude that limiting scientific evidence to quantitative evidence has virtually nothing to do with the natural sciences, but became, in line with an economically driven commodification of knowledge, a dogma of modern social scientists, put forward with a missionary zeal—"Go out and make all humankind measurable!"

The dominance of the quantitative imperative, linking science to quantification, is today not primarily an internal scientific or philosophical problem, but an economical and political issue, and needs be addressed as such. As researchers working with qualitative approaches, we may attempt: (1) not to get drawn into endless arguments about the essences of quantitative and qualitative methods in philosophical or scientific contexts; (2) to adopt the practical approach of natural scientists, and let subject matter and research purpose decide the application of qualitative and quantitative approaches in an investigation; and (3) to address the economic and sociopolitical demands for quantification of human behavior by funding agencies and requirements of evidence-based human practices, where quantified evidence is placed at the top of a knowledge hierarchy, and qualitative evidence at the bottom. Thus, understanding knowledge produced by interviews is not just an epistemological or ontological issue, but also an issue of understanding the prevailing social, cultural, and economical conditions, which find expression in the quantitative imperative. Next, we turn directly to these conditions, considering interviewing as a social practice.

RESEARCH INTERVIEWING AS SOCIAL PRACTICE

Understanding research interviewing involves more than mastering the skills of a craft and the conceptions of knowledge. Interviewing is a social practice; it is a development of everyday conversations. As such, it is a specific form of knowledge seeking, which brings with it specific moral issues. Understanding interviewing also means to understand the social practice of interviewing and how this is influenced by, and itself influences, the larger social situation in which it takes place.

We will here bring together some aspects of the social practice of interviewing from the preceding chapters. This concerns the impact of broader social forces, as well as the more local institutional forces influencing the position of research interviewing. And it concerns the ethical role of research interviewing in providing techniques for controlling human behavior, as well as in shaping academic and public concepts of human behavior.

Research Interviewing in a Social Context

The social and the scientific position of interviewing is historically situated. As a general social practice, interviewing is a new practice from the last few centuries. The confession has a long history as a religious practice, but journalistic, therapeutic, and research interviews proper are rather recent additions to the range of conversational practices in the world's cultures. The interview has, however, been an eminently successful practice, making its way into numerous sectors of modern societies (as illustrated in Box 1.5). Our society has been depicted as an interview society, and the extensive social practice of interviewing provides opportunities and also challenges for current interview researchers. The opportunities, as we have argued above, relate to the idea that interviewing is a highly relevant practice for throwing light on our conversational reality, brought to the extreme in the contemporary interview society. The interview is a natural pathway into the interview society, where people (at least the majority of middle-class citizens) are able and willing to talk about themselves, to reveal or construct aspects of their experiences, personalities, and selves through conversational practices. The other side of this is the challenge to interview researchers that consists of avoiding an uncritical and unknowing reinforcement of central, and possibly problematic, practices of the postmodern consumer society.

Although research interviewing is part of a general social practice in the interview society, as an academic practice, it is a specific, locally determined practice. The position of qualitative research interviewing varies greatly, according to discipline, institutions, countries, and for different periods. Whereas much of what is written in this book may appear old hat and overly meticulous to anthropologists, in some departments of psychology and economy it may appear anarchistic and unscientific. As for our own field—psychology—qualitative research became a legitimate research method in the 1970s in Denmark, in the 1980s in the other Scandinavian countries and in the United Kingdom, and is now on the verge of becoming legitimate in the United States. In the social sciences in general, there has been a strong qualitative movement since the 1970s, which today, in particular in the United States, is threatened by the methodological positivism institutionalized in the bureaucratic requirements of evidence-based practice and the institutional review boards for ethical research practice. We are here faced with the paradox that the scientists' freedom in choosing research methods appropriate to their subject matter is threatened in many academic settings, whereas in the free market of management and marketing, researchers are, as long as their research is in line with company goals, free to choose the appropriate research methods.

Interviewing is a powerful way of producing knowledge of the human situation, and a way that is not ethically neutral. Knowledge is power. The social practice of research interviewing may become a form of radical democratic practice that can be used to help create a free democratic society (Denzin & Lincoln, 2005). However, the social practice of interviewing may also serve to enhance the power of management to control workers and consumers. Power is knowledge. The institutional power to decide which research methods are permissible and deserve funding in an institutional setting also decides which forms of knowledge may be produced. Qualitative research interviews give voice to people in expressing their opinions, hopes, and worries in their own words. Again, we here encounter the paradox that while commercial researchers are free to listen to the voices of the people, these voices may, through the reign of evidence-based practice and the ethical review boards, become out of bounds for academic social scientists.

Returning from the confines of the social practice of research interviewing in academic settings, we find that research interviewing is in line with the general interview culture, a culture that itself contributes to shaping. In this regard, we may ask if "the cult of interviewing" is a reflection of the general "cult of the self," as posited by Atkinson and Silverman (1997). Thus, in *The Fall of Public Man*, Sennett warned against seeing society as a grand, psychological system (Sennett, 1977/2003, p. 4), where the question "Who am I?" is constantly pursued, and where political and other social questions are reduced to a matter of trust, empathy, warmth, and a disclosure of private opinions. Under the conditions Sennett describes as "the tyranny of intimacy," social phenomena are transformed into questions of personality, biography, and individual narratives. The emphasis in much research interviewing on close human interaction in the interview situation should not lead to a neglect of the broader social context of practicing interviewing. It thus becomes pertinent for social science interview researchers to reflect on how their knowledge producing practices relate to and possibly reinforce a tyranny of intimacy (Brinkmann, 2007a).

Interview Ethics in a Social Context

Ethical issues in interview research tend to be raised in relation to the personal implications for the subjects, whereas the wider social consequences of the interviews have received less attention. In line with the common treatments of research ethics, we focused in Chapter 4 on the microethics of the interview situation and on possible future consequences for the subjects involved. We will now go on to draw in a macro-ethical perspective and address potential sociopolitical consequences of the knowledge produced by interviews in a broader social situation (see Brinkmann & Kvale, 2005).

Ethical issues may differ when viewed from a micro and a macro perspective. The respondent may experience an interview situation positively when a researcher with professional authority shows a strong interest in what he or she has to say. The wider social consequences of the knowledge produced in such interviews may, however, be problematic in some cases. This is clear from the Hawthorne studies conducted by management to learn how to manage the workers more efficiently and increase their output, and today in particular in interviewing with the purpose of increasing consumption. Consumer interviews as individual motivational interviews or as focus groups may well follow standard ethical guidelines and also be enjoyable to the participants. On a macro level, however, the consequences are more questionable. Focus group interviews about teenager attitudes to smoking may provide knowledge for improving advertisements to teenagers for smoking, or the knowledge produced may be used in health campaigns to discourage smoking. In a capitalist consumer society it is likely that there will be more capital available for producing and applying knowledge on smoking attitudes for the tobacco industry's advertisements to increase tobacco consumption than for public campaigns seeking to reduce the use of tobacco.

Tensions of ethics on a micro and a macro level also arise in academic interview research. We may here draw in a historic study on anti-Semitism— *The authoritarian personality* by Adorno and his colleagues (1950). In the wake of the Second World War the researchers investigated a possible relation of anti-Semitism to an authoritarian upbringing. The study, which originated from psychoanalytic interviews, used a sophisticated interplay of open qualitative interviews and highly structured questionnaires for producing and validating data. An important part of the investigation consisted of psychoanalytically inspired interviews, where the researchers used therapeutic techniques to circumvent their subjects' defenses in order to learn about their prejudices and authoritarian personality traits. On a micro level, this research clearly violated the ethical principle of informed consent, whereas on a macro level the knowledge obtained of the roots of anti-Semitism was intended to have beneficial social and political consequences in the battle against racial and religious prejudices.

Ethical issues on a macro level can ideally be approached by public discussion of the social consequences and uses of the knowledge produced. We may here again draw in the interview study by Bellah and colleagues about individualism and commitment in America. The researchers saw the very aim of doing social science as a public philosophy, as engaging in debate with the public about the goals and values of society:

> When data from such interviews are well presented, they stimulate the reader to enter the conversation, to argue with what is being said. Curiously, such interviews stimulate something that could be called public opinion, opinion tested in the arena of open discussion. (Bellah et al., 1985, p. 305)

Seeing interviewing as a social practice encourages researchers to go beyond the micro context of the interview situation to conceptualize the effects of interviews on the broader cultural context. Qualitative interviewing as a social practice is at once affected by its social context and can contribute to shaping, supporting, or changing its social context. In other words: We exist in a conversational circle, where our understanding of the social world depends on conversations and our understanding of conversation is based on our understanding of the social world. This is not a vicious circle, but, in a hermeneutical sense, a *circulus fructuosis*. The problem is not to get out of the conversational circle but to get into it the right way.

We may conceive of this task as mastering a "third order" or "triple" hermeneutics. The first order hermeneutics consists of the interview subjects' own understanding of their conversational reality, which already exists in their lived self-interpretations. The research interviewer then undertakes a second hermeneutical interpretation of the interviewees' first order meanings as expressed in the interview conversations. Giddens (1976) has referred to this interpretive process of understanding the actors' own interpretations as a "double hermeneutics." We may continue the hermeneutical move to a third level: If the interviewer returns his or her second order interpretations to the interviewees, the interpretations may, through a third hermeneutical move, enter into and modify the subjects' original interpretations of their activities. And if the interviewer's interpretations are further reported in the public conversation to society at large, they may potentially change the public's understanding of itself, institutions, practices, and people's everyday world. We thus come full circle in a triple hermeneutical process. To give an obvious example, interpretations made from psychoanalytic interviews have today, to a certain extent, become part of Western subjects' understanding of themselves.

Public effects of interview research again draw in ethical conversations about whether the outcomes have been beneficial and to whom. How shall we evaluate ethically the massive influence of psychoanalysis on the public understanding of childhood, personality, and sexuality? Have psychoanalytic qualitative studies contributed to a "triumph of the therapeutic," giving rise to a self-absorbed "psychological man" who is interested in well-being rather than virtue (Rieff, 1966)? Or does psychoanalytic inquiry as a discipline of self-reflection provide a model for developing the emancipatory potentials of social science (Habermas, 1971)? One arena for raising such questions may be to follow the lead of Bellah and his colleagues to use interviews to enhance public conversation. Interviews may, when critically carried out and presented well, incite the reader to enter the conversation and argue with what is said, stimulating a public opinion tested in the arena of public discussion. At the same time, however, interviewers should be aware that not all questions in qualitative social science can be answered through the use of interviews. As we have argued, the object determines the method, and if the object is social interaction in specific practices, for example, fieldwork or participant observation in these practices will likely give more knowledge of the object of interest.

Returning to the issue of qualitative ethicism that we addressed in Chapter 4, we find that a danger in some of today's qualitative literature is a tacit and unquestioned link between qualitative methodologies and an emancipatory agenda. In light of both historical and contemporary qualitative studies, we believe that there is no necessary link between qualitative interview research and a clear ethical conscience. Like all other forms of research, qualitative research holds ethical and political promises—to be able to disclose unacknowledged inequalities and racism, give voice to the oppressed, assist in developing communities in positive directions, and simply enable us to understand our world better—but it also includes ethical and political dangers (e.g., by contributing to the manipulation techniques in management and consumption practices, and we may add in recent years also the political use of focus groups).

Qualitative research interviewing could improve as a knowledge producing social practice by not just being reflexive about what individual researchers bring to the research process (which, for example, may be disclosed through self-analysis), but also by being reflexive about what this social practice of knowledge production—interviewing—as such presupposes about the human reality, and how it affects human reality. Not just self-analysis but also social practice-analysis is required. Reflexivity about the social practice of producing knowledge through interviews should be an intrinsic part of the craft of research interviewing. If done well, by skilled interviewer craftspersons, interview research carries unique potentials for human beings to learn about their meaningful social world.

APPENDIX

Learning Tasks

———◆•◆•◆———

The following learning tasks are for applying, exploring, and extending your knowledge of qualitative research interviewing. The conceptual tasks for the first five chapters of the book serve to stimulate reflection on the nature of research interviewing. For the remaining chapters, there are specific tasks to be carried out for the seven stages of an interview inquiry, which promote the learning of interviewing by doing interviewing. When feasible, carry out the tasks with co-learners. The tasks may also serve as point of departure for group work in class.

The tasks for the first five chapters presuppose that the respective chapters have been read before the tasks are undertaken. For the chapters on interview practice, readers who want to learn the interview craft in ways approximating apprenticeship learning should perform the suggested task first, and thereafter read the chapter. Learning interview research then takes place through one's own practice, and the subsequent reading of the pertinent book chapter may stimulate reflection on one's own learning as well as provide broader perspectives on the tasks undertaken.

CONCEPTUAL TASKS

Chapter 1

Check general textbooks in your own discipline with respect to their references to qualitative interview studies. What are their thematic contributions to the discipline? And how is interview research treated in textbooks on method in your discipline?

Chapter 2

Select a specific form of professional interviewing—for example, a journalist interview, a job selection interview, a legal interrogation, or an oral examination—and try to describe its specific mode of understanding. Then compare your description to the mode of understanding of the qualitative research interview as depicted in Box 2.2.

Chapter 3

Find a book or article that reports interview-based knowledge. What are the implicit or explicit assumptions in the text about knowledge? Does the text represent knowledge or merely subjective opinion about its topic—and why?

Chapter 4

Discuss the following example with respect to ethical guidelines, as well as its broader ethical implications: After you have finished an interview with someone about how she experiences her work and colleagues, and switched off the sound recorder, you talk informally with the interviewee who then gives you new information that makes you see the interview and the interviewee's colleagues in a new light. After the interview, the interviewee contradicts some of her earlier statements and criticizes her colleagues.

Can you use this information in your research—and, if so, how? Should the new information affect your subsequent interviews with the person's colleagues?

Chapter 5

Carry out the transcription task from the first page of Chapter 5. In Box 5.2, potential learning lessons from this task are suggested. Suggesting corresponding lessons from the many learning tasks put forth here would lead too far; for these tasks, readers have to make judgments themselves as to the potential learning benefits and later compare them with the respective parts of the book.

PRACTICAL TASKS

Chapter 6

1. Envisage an interview inquiry, one you are conducting or planning, or just imagine a topic you would like to interview about. Put down on one page of paper an outline of the entire design from original idea to communication of findings. Then compare your overview to the stages depicted in Box 6.2.

2. Ask a fellow student to use your design outline for interviewing you about your experience with, interests in, conceptions of, and prejudices about the topic of your study.

Chapter 7

1. Do a pilot interview of about half an hour. Listen to the recorded interview several times—first with an attitude approximating a free-floating attention, not listening for anything in particular. Then make replays where you focus on specific aspects, such as: What new knowledge was produced during the interview? How was the interaction during the interview? How could your interviewing be improved?

2a. Select and videotape a few interesting TV interviews. Watch them several times and attempt to describe their characteristic styles of interviewing. Note their different forms of questioning, use of pauses, body language, and the like.

2b. Watch another TV interview, first watch it with the sound off, and take notice of the body language. Select still another interview, listen to it with your back to the screen and make an analysis of the interview based only on the sound recording. Then turn around and reanalyze the interview with both the words and the bodies of the speaker present.

Chapter 8

Formulate a research question, then select two or more of the interview forms presented—life world, factual, conceptual, narrative, discursive, confrontational, and focus group—and script interview guides for answering the research question for the interview forms selected.

If feasible, carry out some of the different forms of interviewing in practice.

Chapter 9

Apply the quality criteria for interviews and for the interviewer in Boxes 9.1 and 9.2 to one or more of the interview examples presented in this book. Are the criteria useful, should they be altered or supplemented?

Chapter 10

1a. *Intersubjective transcriber-reliability:* Transcribe a one-page sequence from a research interview. Ask a colleague to transcribe the same page. Then compare the transcriptions and discuss the discrepancies and ways of resolving such divergences in a large research project.

1b. *Intrasubjective transcriber-reliability:* Transcribe a one-page sequence from a research interview, lay it aside for a month, and then retranscribe the same sequence. Compare the transcriptions, reflect on the discrepancies, and consider ways of resolving such differences in a large research project.

2. *Intermodality:* Videotape a research interview and select a sequence corresponding to about one page of transcription. Analyze the video sequence with respect to the research questions underlying the interview. Then try, preferably with co-learners, to analyze the sequence as given in some of the following modalities: the participants' memory of the sequence, the sound recording, the video recording, and a transcribed page. Reflect on potential differences in the analysis and on the strengths and weakness of the different modalities for representing an interview sequence.

Chapter 12

The following interview took place in 1978. The high school teacher was a man in his 50s and the interviewer was a male psychology student.

Interviewer: Has it crossed your mind that grading means something to the pupils' relationship to one another?

Teacher: I haven't noticed anything of that kind. Nothing at all. But it cannot be denied that pupils of course argue with one another, but—is what you ask whether hostility arises among the pupils?

Interviewer: Yes, for instance.

Teacher: Or with knives out and all such things?

Interviewer: Well, whether the marking might lead to either competition or cooperation?

Teacher: I have not at any time noticed that it would lead to, contribute to, any harmful competition. It may now and then happen—when I according to the pupil's opinion mark too low—that they say, "I really think that I am just as good at it as that person there, and he got eight [B], whereas I only got seven [C]." But I don't feel that he or she is trying to say that the other person should have seven.

1. Do the Danish high school teacher's statements confirm the hypothesis that grades further competition among the pupils? Why—or why not?

2. Analyze the meaning of grades expressed by the teacher.

3. Discuss the social interaction in the interview sequence.

Chapter 13

1. The phenomenological interview on learning in Chapter 2 was analyzed in Chapter 12 through a phenomenologically inspired meaning condensation. Undertake a narrative reanalysis of this interview.

2. Bourdieu's interview in Box 1.3 was presented as self-explanatory. Conduct a discursive analysis of this interview.

Chapter 14

Read and analyze the phenomenological interview on learning that appears in Chapter 2 from the perspectives of different theories of learning.

Chapter 15

1. Return to the meaning analysis you have done of the teacher quote for Chapter 12 and discuss how to validate your analysis.

2. Return to Giorgi's meaning condensation in Chapter 12 of the phenomenological interview and discuss the validity of this condensation.

3. Return to Bourdieu's interview in Box 1.3 and discuss the validity of this interview.

Chapter 16

1. Take the three-page sequence from the class interview on grading in Chapter 7 and experiment with different forms of presenting the interview and the main results within an article of not more than three pages.

2. With inspiration from the collage presentation of their interviews by Lather and Smithies in Chapter 16, try to envisage a similar collage presentation of an interview study you are conducting or planning.

GLOSSARY

Audience validation   The presentation of the researcher's interpretations to a lay public for discussion of their validity

Bricolage   Mixed technical and conceptual discourses where the interpreter moves freely between different analytic techniques and theories

Categorization   A systematic coding of a statement into given categories, allowing for quantification

Coding   Breaking a text down into manageable segments and attaching one or more keywords to a text segment in order to permit later retrieval of the segment

Communicative validity   Testing the validity of knowledge claims in a conversation

Communitarianism   A movement in social ethics that rejects the autonomous, liberal self, and instead conceives of the self as constituted by attachments within communities and traditions

Confidentiality   The implication that private data identifying research subjects will not be reported

Confrontational interview   An interview in which the interviewer actively confronts and challenges the views of the interviewee

Consequentialism   The doctrine that the moral rightness of an act is determined solely by the goodness of the act's consequences

Conversation   An oral exchange of observations, opinions, and ideas

Conversation analysis   A meticulous method for studying talk in interaction; it is about what words and sentences do, and the meaning of a statement is understood as the role it plays in a specific social practice

Craft   An occupation requiring special skills and personal know-how, developed through training and long practice

Deconstruction   The destruction of one understanding of a text and opening for reconstruction of other understandings

Deliberate naïveté   The interviewer's exhibiting openness to new and unexpected phenomena, rather than having readymade categories and schemes of interpretation

Dialogue   A conversation between egalitarian partners, who seek true knowledge through argumentation

Discourse analysis   Analysis of the interaction within discourses, on how the talk is constructed, and on what the power effects are of different discursive presentations of a topic within a broad context

Discursive interviewing   An active form of interviewing focusing on the linguistic and social interaction in the interview situation

Doxa   The Greek word for belief or opinion, which does not qualify as genuine knowledge (episteme)

Elite interviews   Interviews with persons who are leaders or experts in a community, usually those in powerful positions

Episteme   Knowledge that has been found to be valid through conversational and dialectical questioning

Epistemology   The study of the nature of knowledge and justification

Ethic of care   An ethic that emphasizes empathy and compassion in personal relationships and a concern with the just community

Ethic of duty   A Kantian ethic of principles, which judges an action according to its intentions, independently of consequences

Focus group interview   A group interview where a moderator seeks to focus the group discussion on specific themes of research interest

Generalizability   The extent that findings in one situation can be transferred to other situations

Grounded theory   A systematic strategy for theory development without a prior theoretical framework; grounded theories are developed through the use of conceptualization to bind facts together, rather than through inferences and hypothesis testing

Hermeneutics The study of interpretations of texts in the humanities; hermeneutical interpretation seeks to arrive at valid interpretations of the meaning of a text

Informed consent Informing the research subjects about the overall purpose and design of an investigation and obtaining their consent to participate

Institutional Review Board (IRB) A board that reviews research proposals with respect to their compliance with ethical guidelines

Intersubjectivity An agreement or understanding between two or more persons about an observation or an argument

Interviewer questions Questions in everyday language that the interviewer poses to the interviewees

Life world The world as it is encountered in everyday life and given in direct and immediate experience, independent of and prior to scientific explanations

Linguistic analysis Analysis that addresses the characteristic uses of language in a speech or text segment, such as the application of grammatical and linguistic forms, the implied speaker and listener positions, and the use of metaphors

Macroethics An ethics that concerns the value and effects of the research-produced knowledge in a sociopolitical context

Meaning condensation An abridgement of the meanings of an interviewee statement into shorter formulations, usually remaining within the understanding and language of the interviewee

Meaning interpretation Interpretation that goes beyond a structuring of the manifest meanings of what is said to deeper and critical interpretations of the text

Member validation The researcher's interpretations presented to the subjects of an inquiry for discussion of their validity

Method A systematic, more or less rule-based, procedure for observation and analysis of data

Methodology The study of the methods of a particular discipline or field

Microethics   An ethics concerning the social interaction in the research situation and the protection of the research subjects

Mixed methods   A combination of different kinds of methods in an investigation, sometimes in the form of paradigmatic contrasts of quantitative and qualitative methods

Narrative analysis   Analysis focusing on the meaning and the linguistic form of texts; it works out plots of interview stories and temporal and social structures

Narrative interviewing   Interviewing in order to elicit or co-construct narratives in the interview, leading to a story with a distinct plot, a social interaction, and a temporal unfolding

Objectivity   The ability to consider or represent facts and information without being biased by personal feelings or opinions

Ontology   The study of "being" and "becoming"; it is concerned with the fundamental nature of existence

Peer validation   The researcher's interpretations presented to peers among researchers for discussion about their validity

Phenomenology   A philosophical perspective based upon careful descriptions and analyses of consciousness, with a focus on the subjects' life world; it attempts to bracket foreknowledge and involves a search for invariant essential meanings of the described phenomena

Phronesis   The intellectual virtue of recognizing and responding to what is most important in a situation; this Greek term may be translated as "prudence" or "practical wisdom"

Positivism   A philosophy that bases science on the observation of data and where the observation of data should be separated from the interpretation of their meanings; truth is to be found by following general rules of method, largely independent of the content and context of the investigation

Postmodernism   A philosophy characterized by a disbelief in modern universal systems of knowledge; it emphasizes the conversational, the narrative, the linguistic, the contextual, and the interrelational nature of knowledge

Pragmatic validity   The responses of the users of an interpretation; in a strong form it concerns whether interventions based on the researcher's interpretations may instigate actual changes in behavior

Pragmatism A move from philosophical legitimation of knowledge to the practical effects of knowledge; knowledge is justified through application, and the strength of our knowledge beliefs is demonstrated by the effectiveness of our actions

Reliability The consistency and trustworthiness of a research account; intra- and intersubjective reliability refer to whether a finding can be replicated at other times and by other researchers using the same method

Research interview A conversation with a structure and a purpose; it involves careful questioning and listening with the purpose of obtaining thoroughly tested knowledge

Research questions The researcher's conceptual and theoretical questions regarding the theme being investigated

Scripting Preparing an interview guide with suggestions for interview questions and their sequence

Second questions The researcher's on-the-spot questions following up an interviewee's answers

Semi-structured life world interview A planned and flexible interview with the purpose of obtaining descriptions of the life world of the interviewee with respect to interpreting the meaning of the described phenomena

Staging Setting the stage for the interview through briefing and debriefing the subjects about the topic and purpose of the interview

Thematizing Explicit formulation of the researcher's conceptualization of the subject matter and the purpose of an investigation

Utilitarianism The moral theory that an action is morally right if and only if it produces at least as much good for all people affected by the action as any alternative action the person could do instead

Validity The strength and soundness of a statement; in the social sciences validity usually means whether a method investigates what it purports to investigate

Virtue ethic An ethic of practical reasoning, after Aristotle, where the personal integrity of the researcher and his or her interaction with the community is crucial

REFERENCES

Adorno, T. W., Frenkel-Brunswik, E., Levinson, D. J., & Sanford, R. N. (1950). *The authoritarian personality.* New York: Norton.

American Psychological Association. (1981). Ethical principles of psychologists. *American Psychologist, 36,* 633–638.

American Psychological Association. (2001). *Publication manual of the American Psychological Association* (5th ed.). Washington, DC: Author.

American Psychological Association. (2002). *Ethical principles of psychologists and code of conduct.* Retrieved June 7, 2007, from http://www.apa.org/ethics/code2002 .html#preamble

Annas, J. (2001). Ethics and morality. In L. C. Becker & C. B. Becker (Eds.), *Encyclopedia of ethics* (2nd ed., pp. 485–487). London: Routledge.

Aristotle. (1994). *Nichomachean ethics.* Cambridge, MA: Harvard University Press.

Atkinson, P., & Silverman, D. (1997). Kundera's immortality: The interview society and the invention of the self. *Qualitative Inquiry, 3,* 304–325.

Atkinson, R. (1998). *The life story interview.* London: Sage.

Barbour, R. (2007). *Doing focus groups.* London: Sage.

Bauman, Z. (1996). From pilgrim to tourist–or a short history of identity. In S. Hall & P. du Gay (Eds.), *Questions of cultural identity* (pp. 18–36). Thousand Oaks, CA: Sage.

Becker, H. S. (1979). Do photographs tell the truth? In T. D. Cook & C. S. Reichardt (Eds.), *Qualitative and quantitative methods in evaluation research* (pp. 99–117). Beverly Hills, CA: Sage.

Bellah, R. N., Madsen, R., Sullivan, W. M., Swidler, A., & Tipton, S. M. (1985). *Habits of the heart: Individualism and commitment in American life.* Berkeley: University of California Press.

Berger, P. L., & Luckmann, T. (1966). *The social construction of reality.* Garden City, NY: Doubleday.

Bernstein, R. J. (1983). *Beyond objectivism and relativism.* Philadelphia: University of Pennsylvania Press.

Birch, M., Miller, T., Mauthner, M., & Jessop, J. (2002). Introduction. In M. Mauthner, M. Birch, J. Jessop, & T. Miller (Eds.), *Ethics in qualitative research* (pp. 1–13). London: Sage.

Bornat, J. (2000). Oral history. In C. Seale, G. Gobo, J. F. Gubrium, & D. Silverman (Eds.), *Qualitative research practice* (pp. 34–47). Thousand Oaks, CA: Sage.

Borum, F., & Enderud, H. (1980). Organisationsforskeren som reporter. Om analyse og rapportering af interview. *Tidsskrift for Samfunnsforskning, 21,* 359–382.

Bourdeau, M., Braunstein, J.-F., & Petit, A. (Eds.). (2003). *Auguste Comte aujourd'hui.* Paris: Kimé.

Bourdieu, P., et al. (1999). *The weight of the world: Social suffering in contemporary society.* Stanford, CA: Stanford University Press.

Bowles, S., & Gintis, H. (1976). *Schooling in capitalist America.* London: Routledge & Kegan Paul.

Brandt, L. W. (1973). The physics of the physicist and the physics of the psychologist. *International Journal of Psychology, 8,* 61–72.

Briggs, C. (1986). *Learning how to ask: A sociolinguistic appraisal of the role of the interview in social science research.* Cambridge, UK: Cambridge University Press.

Brinkmann, S. (2004). Psychology as a moral science: Aspects of John Dewey's psychology. *History of the Human Sciences, 17,* 1–28.

Brinkmann, S. (2006). *Psychology as a moral science.* Unpublished doctoral dissertation, University of Aarhus, Department of Psychology, Aarhus, Denmark.

Brinkmann, S. (2007a). Could interviews be epistemic? An alternative to qualitative opinion polling. *Qualitative Inquiry, 13,* 1116–1138.

Brinkmann, S. (2007b). The good qualitative researcher. *Qualitative Research in Psychology, 4,* 127–144.

Brinkmann, S., & Kvale, S. (2005). Confronting the ethics of qualitative research. *Journal of Constructivist Psychology, 18,* 157–181.

Bruner, J. (1991). The narrative construction of reality. *Critical Inquiry, 18,* 1–21.

Burman, E. (1997). Minding the gap: Positivism, psychology, and the politics of qualitative methods. *Journal of Social Issues, 53,* 785–801.

Cannella, G. S., & Lincoln, Y. S. (Eds.). (2007). Special issue on predatory vs. dialogic ethics. *Qualitative Inquiry, 13,* 315–444.

Charmaz, K. (2005). Grounded theory in the 21st century: Applications for advancing social justice studies. In N. K. Denzin & Y. S. Lincoln (Eds.), *The SAGE handbook of qualitative research* (3rd ed., pp. 507–535). Thousand Oaks, CA: Sage.

Cherryholmes, C. H. (1988). *Power and criticism: Poststructural investigations in education.* New York: Teachers College Press.

Chrzanowska, J. (2002). *Interviewing groups and individuals in qualitative market research.* Thousand Oaks, CA: Sage.

Comte, A. (1975). *Auguste Comte and positivism: The essential writings.* (G. Lenzzer, Ed.). Chicago: University of Chicago Press.

Couper, M. P., & Hansen, S. E. (2002). Computer-assisted interviewing. In J. F. Gubrium & J. A. Holstein (Eds.), *Handbook of interview research* (pp. 557–575). Thousand Oaks, CA: Sage.

Cronbach, L. J. (1971). Test validation. In R. L. Thorndike (Ed.), *Educational measurement* (pp. 442–507). Washington, DC: American Council of Education.

Cronbach, L. J. (1980). Validity on parole: How can we go straight? *New Directions for Testing and Measurement, 5,* 99–108.

Cronbach, L. J., & Meehl, P. J. (1955). Construct validity in psychological tests. *Psychological Bulletin, 52,* 281–302.

Dale, P. A. (1989). *In pursuit of a scientific culture: Science, art, and society in the victorian age.* Madison: University of Wisconsin Press.

Darling-Hammond, L., Wise, A. E., & Pease, S. R. (1986). Teacher evaluation in an organizational context. In E. R. House (Ed.), *New directions in educational evaluation* (pp. 203–253). London: Falmer Press.

Dean, J. P., & Whyte, W. F. (1969). How do you know if the informant is telling the truth? In G. J. McCall & J. L. Simmons (Eds.), *Issues in participant observation* (pp. 105–115). London: Addison-Wesley.

Denzin, N. K. (2001). The reflexive interview and a performative social science. *Qualitative Research, 1,* 23–46.

Denzin, N. K., & Giardina, M. D. (Eds.). (2006). *Qualitative inquiry and the conservative challenge.* Walnut Creek, CA: Left Coast Press.

Denzin, N. K., & Lincoln, Y. S. (2005). Preface. In N. K. Denzin & Y. S. Lincoln (Eds), *The SAGE handbook of qualitative research* (3rd ed., pp. ix–xix). Thousand Oaks, CA: Sage.

Denzin, N. K., & Lincoln, Y. S. (Eds.). (2005). *The SAGE handbook of qualitative research* (3rd ed.). Thousand Oaks, CA: Sage.

Dewey, J. (1930). *Human nature and conduct: An introduction to social psychology.* New York: Modern Library. (Original work published 1922)

Dewey, J. (1958). *Experience and nature.* Chicago: Open Court. (Original work published 1925)

Dewey, J. (1960). *The quest for certainty.* New York: Capricorn Books. (Original work published 1929)

Dichter, E. (1960). *The strategy of desire.* Garden City, NY: Doubleday.

Dinkins, C. S. (2005). Shared inquiry: Socratic-hermeneutic interpre-viewing. In P. Ironside (Ed.), *Beyond method: Philosophical conversations in healthcare research and scholarship* (pp. 111–147). Madison: University of Wisconsin Press.

Donmoyer, R. (1990). Generalizability and the single-case study. In E. W. Eisner & A. Peshkin (Eds.), *Qualitative inquiry in education* (pp. 175–200). New York: Teachers College Press.

Douglas, J. D. (1985). *Creative interviewing.* Beverly Hills, CA: Sage.

Duncombe, J., & Jessop, J. (2002). "Doing rapport" and the ethics of "faking friendship." In M. Mauthner, M. Birch, J. Jessop, & T. Miller (Eds.), *Ethics in qualitative research* (pp. 107–122). London: Sage.

Dreyfus, H. L., & Dreyfus, S. E. (1986). *Mind over machine.* New York: Free Press.

Dreyfus, H. L., & Dreyfus, S. E. (1990). What is morality? A phenomenological account of the development of ethical expertise. In D. Rasmussen (Ed.), *Universalism vs. communitarianism: Contemporary debates in ethics* (pp. 237–263). Cambridge: MIT Press.

Eder, D., & Fingerson, L. (2002). Interviewing children and adolescents. In J. F. Gubrium & J. A. Holstein (Eds.), *Handbook of interview research* (pp. 181–201). Thousand Oaks, CA: Sage.

Eisner, E. W. (1991). *The enlightened eye.* New York: Macmillan.

Eisner, E. W., & Peshkin, A. (Eds.). (1990). *Qualitative inquiry in education.* New York: Teachers College Press.

Ellis, C., & Berger, L. (2003). Their story/my story/our story: Including the researcher's experience in interview research. In J. A. Holstein & J. F. Gubrium (Eds.), *Inside interviewing: New lenses, new concerns* (pp. 467–493). Thousand Oaks, CA: Sage.

Elmholdt, C. (2006). Cyberspace alternativer til ansigt-til-ansigt interviewet. *Tidsskrift for kvalitativ metodeudvikling, 41,* 70–80.

Elster, J. (1980). Metode. In *PaxLeksikon* (Vol. 4). Oslo: Pax.

Eysenck, H. (1973). *The measurement of intelligence.* Lancaster, UK: Medical and Technical Publishing.

Feyerabend, P. (1975). *Against method.* London: Verso.

Fielding, N. (Ed.). (2003). *Interviewing* (Vols. 1–4). Thousand Oaks, CA: Sage.

Fischer, C., & Wertz, F. (1979). Empirical phenomenological analyses of being criminally victimized. In A. Giorgi, R. Knowles, & D. L. Smith (Eds.), *Duquesne studies in phenomenological psychology* (Vol. 3, pp. 135–158). Pittsburgh, PA: Duquesne University Press.

Fisher, S., & Greenberg, R. P. (1977). *The scientific credibility of Freud's theories and therapy.* New York: Basic Books.

Flick, U. (2007). *Designing qualitative research.* London: Sage.

Flyvbjerg, B. (2001). *Making social science matter: Why social inquiry fails and how it can succeed again.* Cambridge, UK: Cambridge University Press.

Flyvbjerg, B. (2006). Five misunderstandings about case-study research. *Qualitative Inquiry, 12,* 219–245.

Fog, J. (2004). *Med samtalen som udgangspunkt* (2nd ed.). Copenhagen: Akademisk Forlag.

Fontana, A., & Prokos, A. H. (2007). *The interview: From formal to postmodern.* Walnut Creek, CA: Left Coast Press.

Foucault, M. (1972). *The archaeology of knowledge.* London: Routledge.

Foucault, M. (1997). On the genealogy of ethics. In P. Rabinow (Ed.), *Essential works of Michel Foucault: Vol. 1. Ethics, subjectivity, and truth* (pp. 253–280). New York: Free Press.

Freud, S. (1963). *Therapy and technique.* New York: Collier.

Frosh, S. (2007). Disintegrating qualitative research. *Theory & Psychology, 17,* 635–653.

Frye, N. (1957). *Anatomy of criticism.* Princeton, NJ: Princeton University Press.

Gadamer, H. G. (1975). *Truth and method.* New York: Seabury Press.

Garfinkel, H. (1967). *Studies in ethnomethodology.* Englewood Cliffs, NJ: Prentice Hall.

Gee, J. P. (2005). *An Introduction to discourse analysis: Theory and method* (2nd ed.). London: Routledge.

Gergen, K. J. (1992). Toward a postmodern psychology. In S. Kvale (Ed.), *Psychology and postmodernism* (pp. 17–30). London: Sage.

Gergen, K. J. (1994). *Realities and relationships. Soundings in social constructionism.* Cambridge, MA: Harvard University Press.

Gibbs, G. (2007). *Analyzing qualitative data.* London: Sage.

Giddens, A. (1976). *New rules of sociological method.* London: Hutchinson.

Giorgi, A. (1970). *Psychology as a human science.* New York: Harper & Row.

Giorgi, A. (1975). An application of phenomenological method in psychology. In A. Giorgi, C. Fischer, & E. Murray (Eds.), *Duquesne studies in phenomenological psychology* (Vol. 2, pp. 82–103). Pittsburgh, PA: Duquesne University Press.

Giorgi, A., & Giorgi, B. (2003). The descriptive phenomenological psychological method. In P. Camic, J. Rhodes, & L. Yardley (Eds.), *Qualitative research in psychology: Expanding perspectives in methodology and design* (pp. 275–297). Washington, DC: American Psychological Association Press.

Glasdam, S. (2007). Interview. En diskussion af ligheder og forskelle i Pierre Bourdieus og Steinar Kvales metodeovervejelser. In K. A. Petersen, S. Glassdam, & V. Lorentzen (Eds.), *Livshistorieforskning og kvalitative interview* (pp. 128–143). Viborg, Denmark: Forlaget PUC.

Glaser, B. G., & Strauss, A. M. (1967). *The discovery of grounded theory: Strategies for qualitative research.* New York: Aldine.

Glesne, C., & Peshkin, A. (1992). *Becoming qualitative researchers: An introduction.* White Plains, NY: Longman.

Gubrium, J. F., & Holstein, J. A. (Eds.). (2002). *Handbook of interview research.* Thousand Oaks, CA: Sage.

Guidelines for the protection of human subjects. (1992). Berkeley: University of California Press.

Habermas, J. (1971). *Knowledge and human interests.* Boston: Beacon Press.

Hammersley, M. (1999). Some reflections on the current state of qualitative research. *Research Intelligence, 70,* 16–18.

Hammersley, M. (2004). Teaching qualitative method: Craft, profession, or bricolage? In C. Seale, G. Gobo, J. F. Gubrium, & D. Silverman (Eds.), *Qualitative research practice* (pp. 549–560). Thousand Oaks, CA: Sage.

Hargreaves, A. (1994). *Changing teachers, changing times: Teachers' work and culture in a postmodern age.* New York: Teachers College Press.

Harré, R. (2004). Staking our claim for qualitative psychology as science. *Qualitative Research in Psychology, 1,* 3–14.

Harré, R., & Moghaddam, F. M. (2003). Introduction: The self and others in traditional psychology and in positioning theory. In R. Harré & F. M. Moghaddam (Eds.), *The self and others: Positioning individuals and groups in personal, political, and cultural contexts* (pp. 1–11). London: Praeger.

Heritage, J. (1984). *Garfinkel and ethnomethodology.* Cambridge, UK: Polity Press.

Hertz, R., & Imber, J. B. (Eds.). (1995). *Studying elites using qualitative methods.* Thousand Oaks, CA: Sage.

Ho, D. Y. F., Ho, R. T. H., & Ng, S. M. (2007). Restoring quality to qualitative research. *Culture & Psychology, 13,* 377–383.

Hollway, W., & Jefferson, T. (2000). Biography, anxiety and the experience of locality. In P. Chamberlayne, J. Bornat, & T. Wengraf (Eds.), *The turn to biographical methods in social science* (pp. 167–180). London: Routledge.

Holstein, J. A., & Gubrium, J. F. (1995). *The active interview.* Thousand Oaks, CA: Sage.

House, E. R. (1980). *Evaluating with validity.* Beverly Hills, CA: Sage.

Howe, R. (2004). A critique of experimentialism. *Qualitative Inquiry, 10,* 42–61.

Hvolbøl, C., & Kristensen, O. S. (1983). Bivirkninger ved karaktergivning. *Psychological Reports Aarhus, 8*(1). Aarhus, Denmark: Aarhus Universitet.

Illouz, E. (2007). *Cold intimacies: The making of emotional capitalism.* London: Polity Press.

Jacobsen, B. (1981). *De højere uddannelser mellem teknologi og humanisme.* Copenhagen: Rhodos.

Jensen, K. B. (1989). Discourses of interviewing: Validating qualitative research findings through textual analysis. In S. Kvale (Ed.), *Issues of validity in qualitative research* (pp. 93–108). Lund, Sweden: Studentlitteratur.

Jonsen, A. R., & Toulmin, S. (1988). *The abuse of casuistry: A history of moral reasoning.* Berkeley: University of California Press.

Karpatschof, B. (2006). *Udforskning i psykologien I–De kvantitative metoder.* Copenhagen: Akademisk Forlag.

Karpatschof, B. (2007). Bringing quality and meaning to quantitative data, bringing quantitative evidence to qualitative observation. *Nordic Psychology, 59,* 191–209.

Keats, D. M. (2000). *Interviewing: A practical guide for students and professionals.* Buckingham, UK: Open University Press.

Kennedy, M. M. (1979). Generalizing from single case studies. *Evaluation Quarterly, 3,* 661–678.

Kerlinger, F. N. (1979). *Behavioral research.* New York: Holt, Rhinehart, & Winston.

Kimmel, A. J. (1988). *Ethics and values in applied social science research.* Newbury Park, CA: Sage.

Kinsey, A. C., Pomeroy, W. B., & Martin, C. E. (1948). *Sexual behavior in the human male.* Philadelphia: Saunders.

Kvale, S. (1972). *Prüfung und Herrschaft.* Weinheim, Germany: Beltz.

Kvale, S. (1980). *Spillet om karakterer i gymnasiet: Elevinterviews om bivirkninger af adgangsbegrænsning.* Copenhagen: Munksgaard.

Kvale, S. (1997). Research apprenticeship. *Nordisk Pedagogik: Journal of Nordic Educational Research, 17,* 186–194.

Kvale, S. (2003). The psychoanalytic interview as inspiration for qualitative research. In P. Camic, J. Rhodes, & L. Yardley (Eds.), *Qualitative research in psychology: Expanding perspectives in methodology and design* (pp. 275–297). Washington, DC: American Psychological Association Press.

Kvale, S. (2006). The dominance through interviews and dialogues. *Qualitative Inquiry, 12,* 480–500.

Labov, W. (1972). *Language in the inner city.* Philadelphia: University of Pennsylvania Press.

Labov, W. (2001, March). *Uncovering the event structure of a narrative.* Paper presented at the Georgetown University Round Table, Washington, DC. Retrieved May 24, 2007, from http://www.ling.upenn.edu/~wlabov/uesn.pdf

Labov, W., & Waletzky, J. (1967). Narrative analysis. In J. Helm (Ed.), *Essays on the verbal and visual arts* (pp. 12–44). Seattle: University of Washington Press.

Laing, R. D. (1962). *The self and others.* London: Tavistock.

Lather, P. (1995). The validity of angels: Interpretive and textual strategies in researching the lives of women with HIV/AIDS. *Qualitative Inquiry, 1,* 41–68.

Lather, P., & Smithies, C. (1997). *Troubling the angels: Women living with HIV/AIDS.* Boulder, CO: Westview Press.

Latour, B. (1997). Foreword: Stengers's shibboleth. In I. Stengers (Ed.), *Power and invention* (pp. vii–xx). Minneapolis: University of Minnesota Press.

Latour, B. (2000). When things strike back: A possible contribution of "science studies" to the social sciences. *British Journal of Sociology, 51,* 107–123.

Latour, B. (2005). *Reassembling the social.* Oxford, UK: Oxford University Press

Lave, J., & Kvale, S. (1995). What is anthropological research? An interview with Jean Lave by Steinar Kvale. *Qualitative Studies in Education, 8,* 219–228.

Lave, J., & Wenger, E. (1991). *Situated learning: Legitimate peripheral participation.* Cambridge, UK: Cambridge University Press.

Levant, R. F. (2005). *Report of the presidential task force on evidence-based practice.* Retrieved February 14, 2008, from http://www.apa.org/practice/ebpreport.pdf

Levine, P. (1998). *Living without philosophy: On narrative, rhetoric, and morality.* Albany: State University of New York Press.

Lewis, O. (1964). *The children of Sanchez.* Harmondsworth, UK: Penguin.

Lincoln, Y. S. (2005). Institutional review boards and methodological conservatism: The challenge to and from phenomenological paradigms. In N. K. Denzin & Y. S. Lincoln (Eds.), *The SAGE handbook of qualitative research* (3rd ed., pp. 165–181). Thousand Oaks, CA: Sage.

Lincoln, Y. S., & Guba, E. (1985). *Naturalistic inquiry.* Beverly Hills, CA: Sage.

Loftus, E. L., & Palmer, J. C. (1974). Reconstruction of automobile destruction: An example of the interaction between language and memory. *Journal of Verbal Learning and Verbal Behavior, 13,* 585–589.

Lovibond, S. (1995). Aristotelian ethics and the "enlargement of thought." In R. Heinaman (Ed.), *Aristotle and moral realism* (pp. 99–120). London: UCL Press.

Løvlie, L. (1993). Of rules, skills, and examples in moral education. *Nordisk Pedagogik, 13,* 76–91.

Lund, N. O. (1990). *Collage architecture.* Berlin: Ernst.

Lyotard, J.-F. (1984). *The postmodern condition: A report on knowledge.* Manchester, UK: Manchester University Press.

MacIntyre, A. (1978). Objectivity in morality and objectivity in science. In H. Tristram Engelhardt, Jr. & D. Callahan (Eds.), *Morals, science and sociality* (pp. 21–39). Hastings-on-Hudson, NY: Institute of Society, Ethics and the Life Sciences.

MacIntyre, A. (2006). *A short history of ethics.* London: Routledge.

Marshall, C., & Rossman, G. B. (2006). *Designing qualitative research.* Thousand Oaks, CA: Sage.

Marx, K. (1998). *The German ideology: Including theses on Feuerbach and introduction to the critique of political economy.* Amherst, NY: Prometheus Books. (Original work published 1888)

Mayo, E. (1933). *The social problems of an industrial civilization.* New York: MacMillan.

Merleau-Ponty, M. (1962). *Phenomenology of perception.* London: Routledge & Kegan Paul.

Michell, J. (2003). The quantitative imperative: Positivism, naïve realism, and the place of qualitative methods in psychology. *Theory & Psychology, 13,* 5–31.

Miles, M. B., & Huberman, A. M. (1984). *Qualitative data analysis: A sourcebook of new methods.* Beverly Hills, CA: Sage.

Miles, M. B., & Huberman, A. M. (1994). *Qualitative data analysis: An expanded sourcebook.* Thousand Oaks, CA: Sage.

Mishler, E. G. (1986). *Research interviewing: Context and narrative.* Cambridge, MA: Harvard University Press.

Mishler, E. G. (1990). Validation in inquiry-guided research: The role of exemplars in narrative studies. *Harvard Educational Review, 60,* 415–442.

Mishler, E. G. (1991). Representing discourse: The rhetoric of transcription. *Journal of Narrative and Life History, 1,* 255–280.

Mishler, E. G. (1999). *Storylines: Craftartist's narratives of identity.* Cambridge, MA: Harvard University Press.

Morse, J. M. (2006). The politics of evidence. In N. K. Denzin & M. D. Giardina (Eds.), *Qualitative inquiry and the conservative challenge* (pp. 79–92). Walnut Creek, CA: Left Coast Press.

Mulhall, S., & Swift, A. (1996). *Liberals and communitarians: An introduction.* Oxford, UK: Blackwell.

Murray, J. A. H., Bradley, H., Craigie, W. A. & Onions, C. T. (Eds.). (1961). *Oxford English Dictionary* (Vol. 5). Oxford, UK: Oxford University Press.

Murray, M. (2003). Narrative psychology and narrative analysis. In P. Camic, J. Rhodes, & L. Yardley (Eds.), *Qualitative research in psychology: Expanding perspectives in methodology and design* (pp. 95–112). Washington, DC: American Psychological Association Press.

National Science Foundation. (2006). *Interpreting the common rule for the protection of human subjects for behavioral and social science research.* Retrieved February 14, 2008, from http://www.nsf.gov/bfa/dias/policy/hsfaqs.jsp#rs

Norris, C. (1987). *Derrida.* London: Fontana Modern Masters.

Nussbaum, M. C. (1986). *The fragility of goodness: Luck and ethics in greek tragedy and philosophy.* Cambridge, UK: Cambridge University Press.

Nussbaum, M. C. (1990). The discernment of perception: An Aristotelian conception of private and public rationality. In *Love's knowledge* (pp. 54–105). Oxford, UK: Oxford University Press.

Ong, W. J. (1982). *Orality and literacy: The technologizing of the word.* London: Methuen.

Palmer, R. E. (1969). *Hermeneutics.* Evanston, IL: Northwestern University Press.

Parker, I. (2005). *Qualitative psychology: Introducing radical research.* Buckingham, UK: Open University Press.

Patton, M. Q. (1980). *Qualitative evaluation methods.* Beverly Hills, CA: Sage.

Pervin, L. A. (1984). *Personality.* New York: Wiley.

Piaget, J. (1930). *The child's conception of the world.* New York: Harcourt, Brace, and World.

Plato. (1953). *V. Lysis, Symposion, Gorgias* (W. R. M. Lamb, Trans.). Cambridge, MA: Harvard University Press.

Plato. (1987). *The republic.* London: Penguin.

Plummer, K. (1995). *Telling sexual stories: Power and change and social worlds.* London: Routledge.

Poland, B. D. (2003). Transcription quality. In J. A. Holstein & J. F. Gubrium (Eds.), *Inside interviewing: New lenses, new concerns* (pp. 267–287). Thousand Oaks, CA: Sage.

Polkinghorne, D. E. (1983). *Methodology for the human sciences.* Albany: SUNY Press.

Polkinghorne, D. E. (1992). Postmodern epistemology of practice. In S. Kvale (Ed.), *Psychology and postmodernism* (pp. 146–165). London: Sage.

Potter, J. (2003). Discourse analysis and discursive psychology. In P. Camic, J. Rhodes, & L. Yardley (Eds.), *Qualitative research in psychology: Expanding perspectives in methodology and design* (pp. 73–94). Washington, DC: American Psychological Association Press.

Potter, J., & Hepburn, A. (2005). Qualitative interviews in psychology: Problems and possibilities. *Qualitative Research in Psychology, 2,* 281–307.

Potter, J., & Wetherell, M. (1987). *Discourse and social psychology.* London: Sage.

Radnitzky, G. (1970). *Contemporary schools of metascience.* Gothenburg, Sweden: Akademiforlaget.

Reason, P. (1994). Three approaches to participatory inquiry. In N. K. Denzin & Y. S. Lincoln (Eds.), *Handbook of qualitative research* (pp. 324–349). Thousand Oaks, CA: Sage.

Reinharz, S., & Chase, S. E. (2002). Interviewing women. In J. F. Gubrium & J. A. Holstein (Eds.), *Handbook of interview research* (pp. 221–238). Thousand Oaks, CA: Sage.

Richardson, L. (1990). *Writing strategies.* Newbury Park, CA: Sage.

Richardson, L., & Adams St. Pierre, E. (2005). Writing: A method of inquiry. In N. K. Denzin & Y. S. Lincoln (Eds.), *The SAGE handbook of qualitative research* (3rd ed., pp. 959–978). Thousand Oaks, CA: Sage.

Ricoeur, P. (1970). *Freud and philosophy: An essay on interpretation.* New Haven, CT: Yale University Press.

Ricoeur, P. (1971). The model of the text: Meaningful action considered as a text. *Social Research, 38,* 529–562.

Rieff, P. (1966). *The triumph of the therapeutic: Uses of faith after Freud*. Chicago: University of Chicago Press.

Roethlisberger, F. J., & Dickson, W. J. (1939). *Management and the worker.* New York: Wiley.

Rogers, C. (1945). The non-directive method as a technique for social research. *The American Journal of Sociology, 50,* 279–283.

Rogers, C. (1956). *Client-centered therapy.* Cambridge, MA: Houghton Mifflin.

Rorty, R. (1979). *Philosophy and the mirror of nature.* Princeton, NJ: Princeton University Press.

Rose, N. (1996). Power and subjectivity: Critical history and psychology. In C. F. Graumann & K. J. Gergen (Eds.), *Historical dimensions of psychological discourse* (pp. 103–124). Cambridge, UK: Cambridge University Press.

Rosenau, P. (1992). *Post-modernism and the social sciences.* Princeton, NJ: Princeton University Press.

Rosenthal, G. (2004). Biographical research. In C. Seale, G. Gobo, J. F. Gubrium, & D. Silverman (Eds.), *Qualitative research practice* (pp. 48–64). Thousand Oaks, CA: Sage.

Rubin, H. J., & Rubin, I. S. (2005). *Qualitative interviewing: The art of hearing data.* Thousand Oaks, CA: Sage.

Runyan, W. M. (1981). Why did van Gogh cut off his ear? The problem of alternative explanations in psychobiography. *Journal of Personality and Social Psychology, 40,* 1070–1077.

Ryen, A. (2002). Cross-cultural interviewing. In J. F. Gubrium & J. A. Holstein (Eds.), *Handbook of interview research* (pp. 335–354). Thousand Oaks, CA: Sage.

Salner, M. (1989). Validity in human science research. In S. Kvale (Ed.), *Issues of validity in qualitative research* (pp. 47–72). Lund, Sweden: Studentlitteratur.

Sartre, J-P. (1963). *The problem of method.* London: Methuen.

Scheflen, A. E. (1978). Susan smiled: On explanations in family therapy. *Family Proceedings, 17,* 59–68.

Scheurich, J. J. (1997). *Research method in the postmodern.* London: Falmer Press.

Schön, D. A. (1987). *The reflective practitioner: How professionals think in action.* New York: Basic Books.

Schön, D. A. (1991). *Educating the reflective practitioner.* San Francisco: Jossey-Bass.

Schwandt, T. A. (2000). Three epistemological stances for qualitative inquiry: Interpretivism, hermeneutics, and social constructionism. In N. K. Denzin & Y. S. Lincoln (Eds.), *Handbook of qualitative research* (pp. 189–213). Thousand Oaks, CA: Sage.

Schwandt, T. A. (2001). *Dictionary of qualitative inquiry.* Thousand Oaks, CA: Sage.

Scientists and thinkers. (1999, March 29). *Time* [Special issue].

Seale, C. (2004). Quality in qualitative research. In C. Seale, G. Gobo, J. F. Gubrium, & D. Silverman (Eds.), *Qualitative research practice* (pp. 407–419). Thousand Oaks, CA: Sage.

Seale, C., Gobo, G., Gubrium, J. F., & Silverman, D. (Eds.). (2004). *Qualitative research practice.* Thousand Oaks, CA: Sage.

Seidman, I. E. (1991). *Interviewing as qualitative research.* New York: Teachers College Press.

Sennett, R. (2003). *The fall of public man.* London: Penguin. (Original edition published 1977)

Sennett, R. (2004). *Respect.* London: Penguin.

Sennett, R. (2006). *The culture of the new capitalism.* New Haven, CT: Yale University Press.

Shaffer, T. L., & Elkins, J. R. (2005). *Legal interviewing and counseling.* St. Paul, MN: West.

Siegel, S. (1956). *Nonparametric statistics for the behavioral sciences.* New York: McGraw-Hill.

Silvester, E. (Ed.). (1993). *The Penguin book of interviews: An anthology from 1859 to the present day.* London: Penguin.

Skårderud, F. (2003). Sh@me in cyberspace. Relationships without faces: The e-media and eating disorders. *European Eating Disorders Review, 11,* 155–169.

Skinner, B. F. (1961). A case history in scientific method. In B. F. Skinner (Ed.), *Cumulative Record* (pp. 76–99). New York: Methuen.

Smith, L. M. (1990). Ethics in qualitative field research: An individual perspective. In E. W. Eisner & A. Peshkin (Eds.), *Qualitative inquiry in education* (pp. 258–276). New York: Teachers College Press.

Spence, D. P. (1982). *Narrative truth and historical truth. Meaning and interpretation in psychoanalysis.* New York: Norton.

Spiegelberg, H. (1960). *The phenomenological movement* (Vol. 2). The Hague, The Netherlands: Martinus Nijhoff.

Spradley, J. (1979). *The ethnographic interview.* New York: Holt, Rinehart & Winston.

Stake, R. E. (2005). Case studies. In N. K. Denzin & Y. S. Lincoln (Eds.), *The SAGE handbook of qualitative research* (3rd ed., pp. 443–466). Thousand Oaks, CA: Sage.

Strauss, A. M., & Corbin, J. (1990). *Basics of qualitative research.* Newbury Park, CA: Sage.

Sullivan, H. S. (1954). *The psychiatric interview.* New York: Norton.

Tanggaard, L. (2007). The research interview as discourses crossing swords. *Qualitative Inquiry, 13,* 160–176.

Taylor, C. (1985). Social theory as practice. In *Philosophical papers: Vol. 2. Philosophy and the human sciences* (pp. 91–115). Cambridge, UK: Cambridge University Press.

ten Have, P. (1999). *Doing conversation analysis.* Thousand Oaks, CA: Sage.

Tesch, R. (1990). *Qualitative research: Analysis types and software tools.* London: Falmer Press.

Tilley, S. A. (2003). Transcription work: Learning though coparticipation in research practices. *Qualitative Studies in Education, 16,* 835–851.

Toulmin, S. (1981). The tyranny of principles. *Hastings Center Report, 11,* 31–39.

van Kaam, A. (1959). Phenomenal analysis: Exemplified by a study of the experience of "really feeling understood." *Journal of Individual Psychology, 15,* 66–72.

Van Maanen, J. (1988). *Tales of the field.* Chicago: University of Chicago Press.

Warren, C. A. B. (2002). Qualitative interviewing. In J. F. Gubrium & J. A. Holstein (Eds.), *Handbook of interview research: Context and method* (pp. 83–101). Thousand Oaks, CA: Sage.

Weitzman, E. A., & Miles, M. B. (1995). *Computer programs for qualitative data analysis.* Thousand Oaks, CA: Sage.

Wengraf, T. (2001). *Qualitative research interviewing.* Thousand Oaks, CA: Sage.

Wetherell, M., & Potter, J. (1992). *Mapping the language of racism: Discourse and the legitimation of exploitation.* Hemel Hempstead, UK: Harvester Wheatsheaf.

Wright Mills, C. (2000). *The sociological imagination.* Oxford, UK: Oxford University Press. (Original edition published 1959)

Yin, R. K. (1994). *Case study research.* Thousand Oaks, CA: Sage.

Yow, V. R. (1994). *Recording oral history.* Thousand Oaks, CA: Sage.

Zuckermann, H. (1972). Interviewing an ultra-elite. *Public Opinion Quarterly, 36,* 159–175.

Zuckermann, H. (1977). *Scientific elite.* New York: Free Press.

INDEX

collage, 288–290
contrasting audiences for,
  267–269
dialogues, 284–285
enhancing, 276–282
enriching, 282–290
ethics and, 272–274
interview quotations in, 279–281
investigating and, 274–276
journalistic interview, 284
metaphors, 287
methodological procedure
  discussion, 278
narratives, 286–287
number of pages in, 282
results in, 278–279
structuring, 277
therapeutic case histories, 285–286
visuals in, 287–288
Interview research:
  ethical issues, 16
  in history and social sciences, 7–12
  quantified knowledge and, 58
Interview sequences, 3–7
Interview situation, ethical
  issues and, 63
Interview society, 12–14
Interview stage, final report and, 275
Interview studies:
  historical, 9–11
  thematizing, 105–109
Interview subjects, 144–147
  across cultures, 144–145
  children, 145–146
  elites, 147
  interview quality and, 165
Interview transcripts:
  analyze vs. narrate, 193
  collected vs. co-authored, 192–193
  living conversations, 192
  method of analysis, 189–192, 194.
    See also Interview analysis
  See also Transcript
Institutional Review Board (IRB), 325
Intrasubjective transcriber-
  reliability, 320

Introductory questions, 135
Investigation of essences, 27

Journalistic interviews, 2, 8, 284

Kantian ethics of duty, 66
Kinsey study, 119–121, 203
Knowledge:
  communication of, 247
  conceptual, of subject matter, 87
  intersubjective, 242–243
  interview. See Interview knowledge
  interviewing as social
    production of, 17–18
  of interviewing, 87
  qualitative, 28, 30
  subject matter, 106–107
  use of term, 2
  See also Interview knowledge
Knowledge collection, 48
Knowledge construction, 48
Knowledge and Human Interests
  (Habermas), 44

Language:
  as medium on interview
    research, 55
  oral and written, 177–178
Leading questions:
  as bias, 301
  children and, 146
  in Bourdieu interview, 159
  in Hamlet, 162–163, 208
  interview quality and, 171–173
  interviewer reliability and, 245
  interviewer results and, 170
  "real meaning" question as, 217
Learning, qualitative research
  interview on, 24–26
Learning How to Ask
  (Briggs), 141
Learning tasks:
  conceptual, 317–318
  practical, 319–322
Life history, 155
Life world, 28, 29, 325

ABOUT THE AUTHORS

————•·•·•————

Steinar Kvale was Professor of Educational Psychology and Director of the Centre of Qualitative Research at the University of Aarhus, and adjunct faculty at Saybrook Institute, San Francisco. He was born in Norway and graduated from the University of Oslo. He continued his studies at the University of Heidelberg with an Alexander von Humboldt scholarship, and has been a visiting professor at Duquesne University, West Georgia University, and the University of Bergen. His long-term concern was with the implications of such continental philosophies as phenomenology, hermeneutics, and dialectics for psychology and education. He studied examinations and grading, and is the author of *Prüfung und Herrschaft* (1972). He edited *Issues of Validity in Qualitative Research* (1989) and *Psychology and Postmodernism* (1992). And with Klaus Nielsen he edited *Mesterlære–Læring som social praksis* (*Apprenticeship: Learning as Social Practice,* 1999) and *Praktikkens lærings-landskab–At lære gennem arbejde* (*The Learning Landscape of Practice: Learning Through Work,* 2003). He died in 2008.

Svend Brinkmann is Assistant Professor of Psychology at the University of Aarhus, Denmark, where he is a member of the Centre of Qualitative Research. He has studied psychology and philosophy at Aarhus and Oxford universities and received his Ph.D. in psychology from Aarhus University for a study of the moral presuppositions and implications of psychology, entitled *Psychology as a Moral Science.* His research interests include qualitative methods, moral inquiry, the philosophy of psychology, and approaches to human science such as pragmatism, hermeneutics, and postmodernism. He is the author of *John Dewey–En introduktion (John Dewey: An Introduction,* 2006) and co-editor with Cecilie Eriksen of *Selvrealisering–Kritiske diskussioner af en grænseløs* *udviklingskultur* (*Self-realization: Critical discussions of a boundless culture of development*, 2005), and with Lene Tanggaard of *Psykologi: Forskning og profession* (*Psychology: Research and Profession*, 2007).